Polymeric Nanoparticles and Microspheres

Editors

Pierre Guiot, Ph.D.
Unité de Chimie Thérapeutique,
University of Louvain, and
International Institute
of Cellular and
Molecular Pathology
Brussels, Belgium

Patrick Couvreur, Ph.D.
Professor
Laboratoire de Pharmacie
Galénique et Biopharmacie
Centre d'Etudes Pharmaceutiques
Université de Paris XI
Chatenay-Malabry, France

CRC

CRC Press, Inc.
Boca Raton, Florida

Library of Congress Cataloging-in-Publication Data
Main entry under title:

Polymeric nanoparticles and microspheres.

Bibliography: p.
Includes index.
1. Nanoparticles. 2. Microspheres (Pharmacy)
3. Drugs—Vehicles. I. Guiot, Pierre, 1947-
II. Couvreur, Patrick, 1950-
RS201.N35P65 1986 615′.191 85-21307
ISBN 0-8493-5696-2

This book represents information obtained from authentic and highly regarded sources. Reprinted material is quoted with permission, and sources are indicated. A wide variety of references are listed. Every reasonable effort has been made to give reliable data and information, but the author and the publisher cannot assume responsibility for the validity of all materials or for the consequences of their use.

Direct all inquiries to CRC Press, Inc., 2000 Corporate Blvd., N.W., Boca Raton, Florida, 33431.

International Standard Book Number 0-8493-5696-2

Library of Congress Card Number 85-21307
Printed in the United States

To
MICHELINE
To
CECILE, CATHERINE, MARIE, and NICOLAS

PREFACE

Paul Ehrlich's image of a "magic bullet" is one answer to the problem of drug action specificity. Though perhaps an idealistic image, it is true that the concept of drug targeting is gaining more and more interest in the medical community. The rationale behind the concept is to bring molecules into intimate contact with an elusive target in a controlled way. Beside specificity, other advantages may be protection against potential premature inactivation and decreased toxicity by using reduced doses.

Historically, the basic philosophy behind the development of drug therapies was to push drug action as much as possible. Indeed, the pharmacological response to a drug is directly related to drug concentration at the required site. Because the body distribution of a molecule is essentially based on physicochemical properties, not necessarily fitting the characteristics of the diseased area, large amounts of drug have to be given. Toxicity comes then from massive drug penetration into healthy organs and tissues. Until now, changing chemical structure of the drug has been the only way to improve the selectivity of an active molecule.

An alternative philosophy is to start at the zero drug level and find the amount which is required by the affected area only. Novel approaches to drug delivery concern the association of an active molecule with a carrier system. Liposomes are certainly the most developed carrier system and have been extensively described in publications, reviews, and books. There has been great interest generated in the use of liposomes for membrane model studies, but, they do present some disadvantages, particularly related to their in vivo instability and to difficulties in their preparation at an industrial level.

Another type of carrier system is solid polymeric nano- or microparticles. These new pharmaceutical formulations are generally able to carry a wide variety of drugs in a more stable and reproducible way. Considering the number of newly published papers dealing with polymeric nanoparticles or microspheres, it is clear that interest in the development of such drug delivery systems is increasing. Nevertheless, as far as we know, there is no general book covering the subject. The aim of this work is to partially fill this lack and we hope it will act as a great stimulus for developing new polymeric drug delivery systems.

Contributors were asked to emphasize their own experience with nanoparticles or microspheres. They were also encouraged to provide aspects normally not stipulated in specialized reviews. The objective was to produce a book which although written by specialists, presented a subject which could be easily understood by people in different fields of life sciences. Another goal was to set up a condensed work covering many aspects of nanoparticles and microspheres. That is the reason why, although mainly dealing with synthetic drug carriers, the first chapter is related to preparation and physicochemical properties of nanoparticles constructed from natural polymers. This is followed by two chapters involving the application of polymeric drug carriers in cancer therapy. The second chapter extensively describes in vitro and in vivo behavior of purely synthetic nanoparticles. As in the case of liposomes, a problem can be that some particles are preferentially taken up by the reticuloendothelial system. Magnetic drug carriers provide a partial answer to this problem since their tropism can be managed by an external magnet. The third chapter is entirely devoted to this aspect. A brief but original application of nanoparticles is given in the fourth chapter by reviewing their use in ocular therapy. Finally, two general applications of polymeric microspheres are discussed in Chapters 5 and 6. These micrometer-sized spherules are of tremendous interest either for embolization or for diagnostics.

The editors thank all contributors for their valuable work and CRC Press Inc. for their professional cooperation.

P. Couvreur and P. Guiot
July 1984

THE EDITORS

Pierre Guiot, Ph.D., obtained an engineering degree from the Meurice Institute of Chemistry. Having performed theoretical studies on the diffusion of spherical particles, he received his M.Sc. in physics from the Free University of Brussels. He then attended the International Institute of Cellular and Molecular Pathology, where he worked on the quantification of the interactions between cells and drug carriers (liposomes, nanoparticles etc.) obtaining a Ph.D. in biophysics. By receiving a prize from the Royal Academy of Sciences of Belgium among other academic honors, he was encouraged to pursue his research. At the Department of Medicinal Chemistry of the University of Louvain he carried out investigations in drug targeting and drug pharmacokinetics. Dr. Guiot is a member of the Belgian Biochemical and Biophysical Societies and also the Chemical Society of Belgium. He is also working in The Netherlands, in the Research and Development Department of Medtronic Europe. There, he contributed to the field of chronopharmacology of chemotherapeutic agents by the use of drug administration devices and systems. Dr. Guiot is now Director of ELA Medical Benelux.

Patrick Couvreur, Ph.D. is Professor of Physical Pharmacy and Biopharmacy at the University of Paris XI, France. Dr. Couvreur obtained his title of Pharmacist from the Catholic University of Louvain in 1972 and his Ph.D. from the same university in 1975. He served as "Chargé de Recherches" at the National Belgian Fonds for Scientific Research from 1977 to 1979. After a stay of 1 year at the Federal Institute of Technology of Zurich, Switzerland, he became Senior Lecturer at the Catholic University of Louvain and obtained the title of "Agrégé de l'Enseignement Superieur" in 1983. He assumed his actual position in 1985.

Professor Couvreur has presented numerous invited papers and lectures at international meetings and has published more than 40 research papers mainly in international journals. Owner of more than 10 patents at the international level, he has written chapters in five books. He received, among others, the Prize Alvarenga 1983 from the Belgian Academy of Medicine.

His current research interests include the development of new drug-carriers and the use of biopolymers in the design of drug-delivery systems.

CONTRIBUTORS

Jean-Pierre Benoit, Ph.D.
Assistant
Pharmacie Galénique et Biopharmacie
Laboratoire de Centre d'Etudes Pharmaceutiques
Université de Paris XI
Chatenay-Malabry, France

Aleksander Biernacki, M.D.
Research Assistant
Laboratoire Cancérologie Expérimentale
Faculty of Medicine
University of Louvain
Brussels, Belgium

Francis Brasseur, Ph.D.
Research Assistant
Laboratoire Cancérologie Expérimentale
Faculty of Medicine
University of Louvain
Brussels, Belgium

Patrick Couvreur, Ph.D.
Professor of Pharmacy
Laboratoire de Pharmacie Galénique et Biopharmacie
Centre d'Etudes
Pharmaceutiques
Université de Paris XI
Chatenay-Malabry, France

S.S. Davis, Ph.D., D.Sc.
Lord Trent Professor of Pharmacy
Department of Pharmacy
University of Nottingham
Nottingham, England

Malcolm Frier, Ph.D.
Principal Radiopharmacist
Department of Radiopharmacy/ Medical Physics
Queens Medical Centre
Nottingham, England

Luc Grislain, Ph.D.
Manager
Biocinétique et Métabolisme
Technologie Servier
Orléans, France

Pierre Guiot, Ph.D.
Chemical Engineer
Director
ELA Medical Benelux
Brussels, Belgium

Robert Gurny, Ph.D.
Lecturer
School of Pharmacy
University of Geneva
Geneva, Switzerland

Lisbeth Illum, Ph.D.
Associate Professor
Department of Pharmaceutics
Royal Danish School of Pharmacy
Copenhagen, Denmark

Vincent Lenaerts, Ph.D.
Manager
Department of Pharmaceutics
Laboratoires UPSA
Rueil-Malmaison, France

Yasunori Morimoto, Ph.D.
Associate Professor
Department of Pharmaceutics
Josai University
Sakado/Saitama, Japan

Richard C. Oppenheim, Ph.D.
Senior Lecturer
School of Pharmaceutics
Victorian College of Pharmacy
Parkville, Victoria, Australia

Francis Puisieux, Ph.D.
Professor
Laboratoire de Pharmacie Galénique et Biopharmacie
Centre d'Etudes Pharmaceutiques
Université de Paris XI
Chatenay-Malabry, France

Kenji Sugibayashi, Ph.D.
Research Associate Professor
Department of Pharmaceutics
Josai University
Sakado/Saitama, Japan

TABLE OF CONTENTS

Chapter 1

NANOPARTICULATE DRUG DELIVERY SYSTEMS BASED ON GELATIN AND ALBUMIN

Richard C. Oppenheim

TABLE OF CONTENTS

I. INTRODUCTION

Until very recently mankind has used innocuous drugs inefficiently. With the chemotherapeutic revolution earlier this century, it became clear that drugs cannot only cause the desired response but they can also have adverse effects on the body.

Traditional pharmaceutics has involved individual pharmacists being concerned with producing an elegant product with little thought needed on its stability, therapeutic index, or drug targeting capacity. Most of the industrial dosage forms in use today are a carry over from these nonscientific times. If we tried to design an efficient delivery system today, it is unlikely that we would produce a product in which the drug is dispersed in a compressed powder bed and then the product administered orally. There are too many ways in which the drug may be prevented from reaching the desired site of action. The in vitro disintegration and dissolution tests imposed by pharmacopoeia and regulatory authorities are attempts to obtain a more reproducible in vitro product. The numbers generated in these tests look nice but without any correlation with in vivo efficacy they only satisfy bureaucratic insecurity. Many measures of in vivo efficacy including clinical response, drug concentration near the site of action, or plasma drug concentration have been proposed. None of these laudable measures are tackling the twin issues facing drug delivery within a living body. For effective therapy we need to know:

1. How much drug is needed at the site of action to cause the desired response
2. How can the drug go to this site of action, and nowhere else, in a form and at a rate to obtain the desired response

To obtain this information for new drugs involves a considerable investment of time, money, and effort. Only large organizations, such as major pharmaceutical companies or government-sponsored research organizations can be expected to have these resources. It is most unlikely that this information will ever be generated for the vast majority of drugs currently approved for marketing, simply because there are neither economic nor regulatory pressures to do so.

As more potent drugs have been introduced, the community and hospital pharmaceutical profession has adjusted to the reduction in doses from grams to milligrams. However, the traditional nature of the delivery system means that the bulk of the administered drug may either cause an adverse response or be metabolically sacrificed so that the remaining small percentage of the dose can do the necessary interaction at the desired site of action.

Most pharmaceutical companies with a realistic approach to drug utilization have recognized the inefficiencies in traditional dose forms and have established novel drug delivery groups. These groups are given the responsibility of devising delivery systems that only need the micro- or picogram quantities used in effective targeted therapy. The groups should consider established line drugs as well as new drugs. They should also draw upon the expertise of smaller academic research groups who are keen to work with industry to produce better therapy.

As these targeted drug delivery systems are being prepared for marketing, the emphasis that has been placed over the last 20 years on correlations based on drug plasma levels, the need to quantify plasma levels, and the requirement to perform black box manipulations on plasma data will be seen as misplaced.

The analytical chemists will have had to develop new strategies to quantify the very small amounts of drug needed at the site of action. It will no longer be appropriate to rely on equilibria between the site of action and general plasma levels.

The formulators have to overcome the problem of using such low doses that only a very small amount of degradation will destroy a significant fraction of the potency.

The temptation to revert to elegant but ineffective products with the drug buried away from attacking degradants, will have to be resisted.

Two different approaches have been developed to improve traditional parenteral administration of drug solutions or suspensions. The local perfusion approach is represented by the embolism within the capillaries between the arterial and venous systems. The degradable particles used have a size from about 10 to 100 μm. However, there is little or no direct interaction of the delivery system with the membrane of the cell containing the target site.

The second approach acknowledges that cells are at the most only a few microns in diameter. These submicron delivery systems are used to interact with the target cells. If the target site is within a cell and the delivery system is taken into the cell and degraded within the cell, the resultant local perfusion is within a much smaller volume, and hence is more likely to be effective than general bathing of the target cell in a wash liquid containing drug.

By analogy to the micron sized microparticles or microspheres commonly used in the embolic approach to targeting, the general term of nanoparticle is used to describe these nanometer-sized particles.[1]

A clear distinction is drawn between the particulate nanoparticles and smaller drug-macromolecular complexes. In such complexes, the macromolecule can be DNA,[2] albumin,[3] polypeptides such as polylysine,[4] polysaccharides,[5] hormones,[6] lectins,[7] or antibodies.[8] This distinction is based on the accessibility of the drug to other molecules which may degrade it. In nanoparticles, the drug is molecularly dispersed within a biodegradable particle and provided the particle is intact, the drug is protected or entrapped within a porous matrix and could encounter degradants only if it diffuses out of the particulate delivery system.

The work reviewed in this chapter involves drug delivery systems which fall into the second approach to improving drug delivery. It focuses on the use of natural macromolecules such as gelatin and albumin as the base for the delivery system. Hence, it is complementary to those chapters which concentrate on synthetic polymers. There are a number of ways in which the proteins used can be fixed into the aggregated state. This chapter concentrates on the desolvation, chemical hardening approach and so is also complementary to the chapter which reviews inter alia, heat aggregation, and emulsification procedures. This chapter gives practical guidelines on how to make such nanoparticles and having achieved an elegant in vitro product, some discussion is made on the use of these nanoparticles.

II. CRITERIA FOR EFFECTIVE NANOPARTICLES

In designing a solid colloidal drug delivery system, various ideal criteria can be proposed for both the base macromolecule and also for the overall delivery system.

The carrier macromolecule should be nontoxic and biodegradable. It is a pointless exercise to develop a drug delivery system using a toxic macromolecule. If the molecule has to be polymerized to form the delivery system, the monomeric form must either be shown to be entirely absent or totally nontoxic. While some synthetic macromolecules used can be criticized on these criteria, natural macromolecules should have no inherent toxicity of their own. It is appropriate, however, to express concern about the immunological response to repeated administration of drug delivery systems based on macromolecules such as albumin or gelatin. Encouraging indications are shown by the routine use of macroaggregated albumin (MAA) systems as diagnostic agents. Many colloidal particles used in radiopharmacy are stabilized by gelatin. In both these situations, there is an extremely low incidence of adverse reactions. Gelatin dispersions have been used as emergency plasma substitutes. The experimental results accumulated so far in this program and summarized later in this chapter have been sufficiently encouraging to enable the program to continue.

The other major criterion of the macromolecule used as the base for the nanoparticle is that it should enable the delivery system to be biodegraded. It is clearly inappropriate to make an elegant delivery system containing a high payload of drug which targets the desired site of action but which cannot release the drug in vivo. The switch from the use of methylmethacrylate to the alkylcyanoacrylates by European research groups is a recognition of the greater in vivo hydrolytic degradation of the latter macromolecules. Nanoparticles made from proteins need unrealistic extremes of pH and very high temperatures before they undergo hydrolytic degradation.[9] For such nanoparticles to break down, an enzymatic mechanism must be invoked. The appropriate degradation procedure has to be available at the site of action.

The time scale of biodegradation is also important. Very early in the development of a drug delivery system, a decision has to be made whether the drug is to be released as the nanoparticle degrades or whether the drug is released by diffusional processes and the nanoparticle degrades over a much longer period. Such a decision is identical to that required for microparticles and implants and has been reviewed in depth many times as for example by Robinson[10] and Chien.[11] The MAA systems are essentially nonporous as are the hardened proteinaceous coatings on microcapsules. It would seem that if albumin and gelatin are to be used as the carrier material in nanoparticles, the drug will probably only be released from the delivery system as that system erodes.

While these are some of the criteria for the carrier part of the drug delivery system, there are also a number of important criteria that the whole delivery system should try to meet.

First, the overall drug delivery system must be pharmaceutically acceptable. A shelf-life for the product which ensures transfer from an industrial manufacturing environment to the clinical setting is most important. The original fluid liposomes with a very short shelflife and stringent storage conditions had no hope of commercial exploitation. While no hard and fast rules can be laid down, a shelflife of at least 1 year when stored at temperatures less than 30°C would seem to be the minimum before any industrial exploitation of the system might be possible. If the product is stored as a wet dispersion, there should be no likelihood of aggregation.[12] Single discrete deflocculated particles will eventually settle out unless they have the same density as the vehicle or they are so small that Brownian motion dominates. Upon settling out, coagulation can occur and the resultant particle cake at the bottom of the container will be extremely hard to redisperse. If the particle volume fraction is sufficient to enable a flocculated formulation, considerable care is needed to ensure that neither syringe needles nor body circulating systems become blocked with clumps of particles.

If the product is stored as a lyophilized powder, the time and extent of redispersion with solvent needs to be well controlled. No aging effects can be allowed to occur in the powder.

Second, although some work has been done on peroral administration[13] most of the nanoparticles are given parenterally. Hence, the product has to be produced in a sterile manner or be subjected to terminal sterilization.

Most nanoparticle products, including the protein-based products, are passed through a size exclusion column to purify them. Provided the column is bacteriologically clean and stable, the void volume is unlikely to be contaminated. Particular care is required with the proteinaceous products since the protein may act as a food source for the contamination. However, starting with high quality material and anticipating that glutaraldehyde will be added during the production of the nanoparticles, viable microorganism and viral contamination in the finished product is expected to be very low. When 1 mg samples of a number of batches of gelatin-based nanoparticles were aseptically placed in 10 mℓ of tryptone soya bean broth and incubated at 37°C for 48 hr, no turbidity developed.[14] This indicates the absence of aerobic bacteria and fungi in a random selection of proteinaceous nanoparticles.

It may seem obvious to research groups who produce nanoparticles that it is inappropriate to terminally sterilize nanoparticles by passage through membrane filters without regard to the size of the nanoparticles. However, there are two cases at least where medical research scientists working in collaboration with but distant from the nanoparticle group who terminally filtered a nanoparticle product and then were surprised to find little activity in the filtrate. Such experiences show that in collaborative studies it is imperative that the producers of the nanoparticles are either present when the in vitro and in vivo testing is done or that an extremely detailed protocol is established and followed exactly to ensure that the drug delivery system has a fair trial.

In the product development stage, sterility is not necessarily a high priority. However, for successful marketing sterility must be assured. Membrane filtration is likely to compromise product yields, irradiation may cause unwanted polymerization or drug degradation, and exposure to gases such as ethylene oxide may lead to residue problems. Hence, sterile manufacture using sterile material would appear to be the only feasible alternative. Proteinaceous nanoparticles are made in an aqueous environment and do not have the advantages of some of the other nanoparticle products which include the use of a nonaqueous solvent in the manufacturing procedure. In consequence, all other things being equal, the proteinaceous nanoparticles are likely to cost more to produce.

If targeting is to be achieved, the drug delivery system must get the payload of drug to the site of action and nowhere else. Any system that has the possibility of desorbing or releasing drug either when the product is being stored or prepared for administration or after the product has been administered and before interaction with the target cell or site of action is not perfect. Any porous, sponge-like system can appear to have a high drug payload.[15] However, the important payload figure is that which cannot be readily removed upon placing the delivery system into a drug-free vehicle.

It is important to know where the desired site of action is located in the target cell. Considerable effort has gone into producing a variety of colloidal delivery systems containing adriamycin. These lysosomotropic systems deliver the cytotoxic into the interior of the target tumor cell. It is now clear[16] that adriamycin also has a site of action on the outer membrane of the target cell. The solid nanoparticle products which are not capable of fusion with the cell membrane may be inferior to the fluid liposomes which, despite all their other shortcomings, at least do have the possibility of membrane fusion.

The final criterion that the delivery system should meet is that the drug must be released from the delivery system at the site of action at an appropriate rate. Using traditional delivery systems, we assume that the resultant circulating drug penetrates all the necessary membranes and arrives at the site of action at a rate suitable for effective therapy. As a general rule, we do not know what is the appropriate rate of arrival or the drug concentration at the site of action. If a targeted system is to be developed, these site of action values need to be established so that there is neither an over nor under supply of drug. Such an undertaking is likely to involve a large multidisciplined team and is almost perforce restricted to industrial novel delivery groups. In most of the diseases for which a nanoparticle system has been proposed, the desired response is death of the cell. Hence, a rapid release of a high payload is appropriate.

III. DESOLVATION OF PROTEINS LEADING TO NANOPARTICLES

Physical chemists for almost 100 years have been studying the processes occurring when taking a solution of a macromolecule and adding a liquid which although misci-

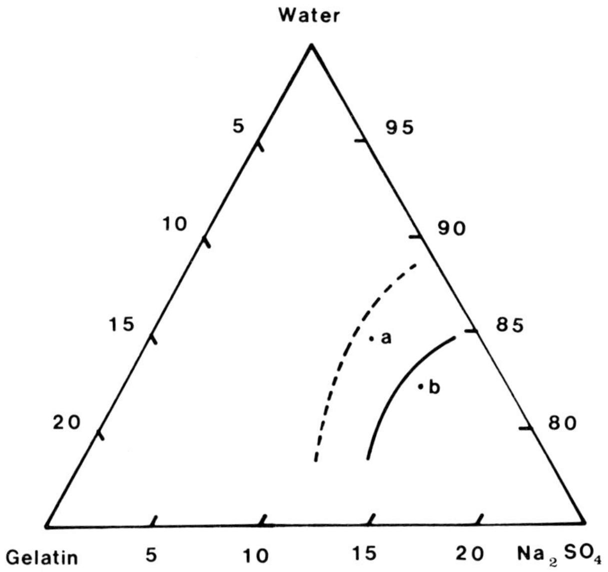

FIGURE 1. Schematic representation of desolvation of a gelatin solution by a sodium sulfate solution; ----- coacervation phase boundary —— precipitation boundary.

ble with the original vehicle is a nonsolvent for the macromolecule. If sufficient liquid is added, the macromolecule becomes desolvated. Two major stages are recognized in this desolvation process and are represented by the two-phase boundary lines in Figure 1. At point a, the system has been desolvated only to such an extent that it separates into two liquid phases. The protein-rich phase, or coacervate, contains an enriched amount of protein in droplets which form a viscous dispersion in the protein deficient background phase. These small, high surface area droplets seek surfaces and interfaces upon which to deposit and coalesce. The coalesced material can then be chemically crosslinked by the addition of aldehydes such as formaldehyde or glutaraldehyde. This extent of desolvation is the basis of the well-known microencapsulation procedure of coacervation and phase separation.

If the desolvation is continued further, to say point b in Figure 1, the protein will precipitate from the very poorly solvating environment. Such procedures are well known in traditional protein fractionation and more recently have become fashionable in antibody purification.[17]

In late 1973, just as a microencapsulation project was commencing,[18] the ongoing link between the Victorian College of Pharmacy and Speiser's group at the ETH in Zurich was being formed. The combination of these events led to the question: what happened if the desolvation of the protein was stopped just outside the coacervate phase boundary and the crosslinking agents added? It was hoped that small discrete particles could be formed although it was recognized that there may be insufficient separation between the macromolecule chains and excessive crosslinking may lead to a solid gelatinous mass being formed.

A simple procedure was needed to monitor the extent of desolvation and the approach towards the coacervate phase boundary. We initially used dark field microscopy of Tyndall light to identify when the coacervation phase boundary was crossed and droplets formed. However, although sensitive to the formation of coacervate droplets, it was difficult to use the procedure to follow the desolvation of the system before the droplets started to form. Nephelometry using a simple white light from a

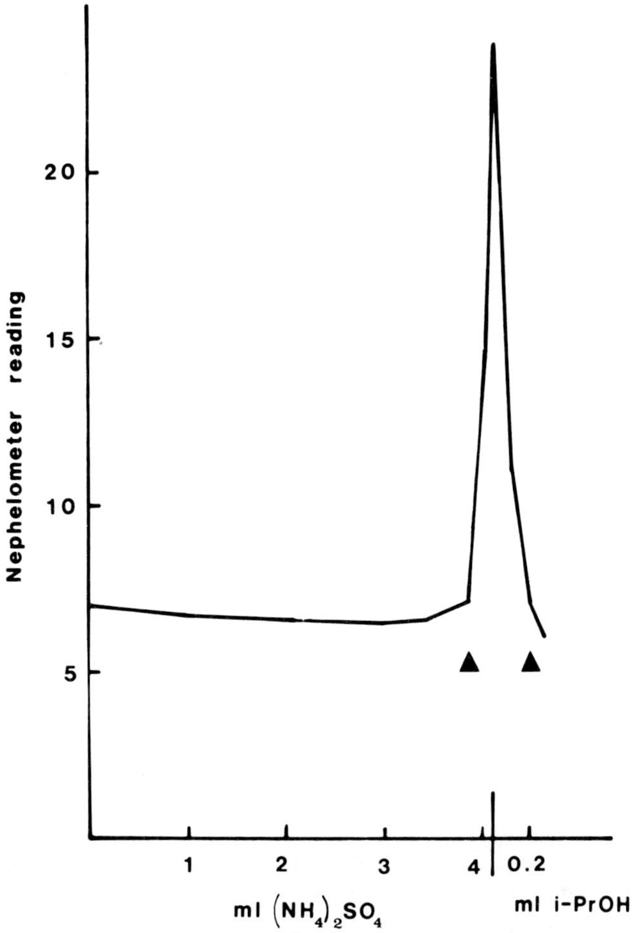

FIGURE 2. Light scattering measured as a bovine serum albumin solution is desolvated with ammonium sulfate solution and resolvated with isopropanol. The ideal points for hardening with glutaraldehyde are indicated by arrows.

light bulb and measured on an inexpensive photoelectric cell was found to give reproducible profiles as the desolvation proceeded. All the data presented in this chapter use a simple nephelometer such as the Turner Model 40 nephelometer. Improvements would include the use of a single wavelength light source such as a laser and to monitor at a variety of angles. In this way the particle size of the forming nanoparticles might be estimated.

Figure 2 gives a typical desolvation profile. As the ammonium sulfate solution is added to this system, the extent of light scattering decreases slightly. However, as the coacervate phase boundary is approached there is a rapid rise in the intensity of the scattered light until the system becomes visibly turbid as coacervation occurs. Resolvation with a small amount of water or low molecular weight alcohols leads to a reduction in the case of water or a temporary reduction for the alcohols in the extent of light scattering as the system crosses the boundary back out of the coacervate state to a more isotropic state.[19] Most of the nanoparticles based on gelatin or albumin have been made when the light scattering-desolvation profile is just starting to rise and these points are indicated on Figure 2.

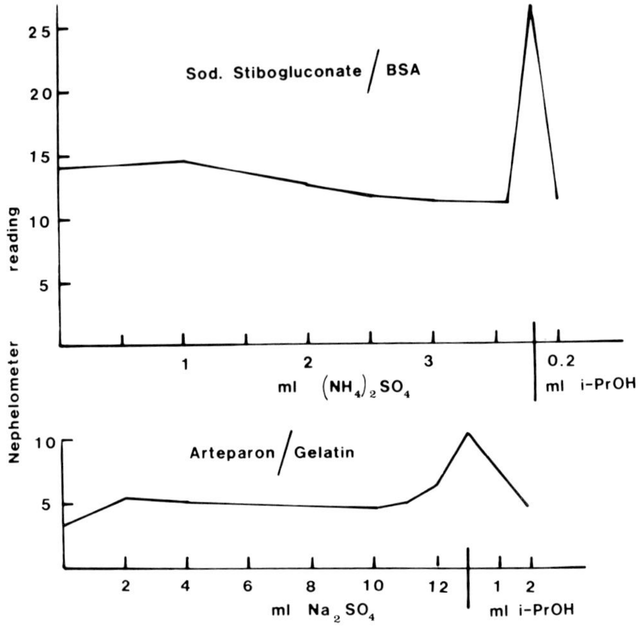

FIGURE 3. Light scattering profiles for BSA (upper) and gelatin (lower) based systems.

On a laboratory scale, it is fairly easy to monitor any profile and with experienced professional operators it is easy to decide when the appropriate point for hardening has been reached. However, for larger scale in-plant production, this point has to be well defined and readily identifiable. If the point is overshot by just a very small addition of desolvating agent and coacervation occurs, the system will have to be resolvated to regain the desired extent of desolvation just outside the coacervate phase boundary. It will be critical that very tight controls be placed on temperature and total system volume as well as amounts of protein and reagents used. Alternatively, if the desolvation profile is flat and the rise towards the coacervated state gentle, then it is difficult to decide just when the appropriate amount of desolvating agent has been added.

Figure 3 shows that for a bovine serum albumin (BSA) based system, the rise is much sharper than for a gelatin system. This typical difference is ascribed to the disperse nature of the gelatin macromolecules having been derived from hydrolysis of collagenous material. The albumins used should have a much more uniform molecular weight and so all the albumin molecules should exhibit uniform desolvation and a sharper approach to the coacervate phase boundary.

Different amounts of desolvating agents need to be added to achieve the desired state ready for hardening into nanoparticles. Different lots of nominally the same gelatin may need different amounts of desolvating agents. This is probably the result of differences in the average molecular weight as described by Nixon et al.[20] Specifications for gelatin need to include a desolvation profile to ensure that batch to batch reproducibility with the same operating procedures can be achieved. We routinely use a 20% sodium sulfate solution to desolvate gelatin-based systems and a 40% ammonium sulfate solution to desolvate albumin-based systems, although the sodium sulfate solution may also be used for the albumin systems. The ammonium sulfate solution acts as a weak buffering agent and unless the pH is monitored and controlled during the desolvation, there is a drift in pH as the desolvation occurs. Different amounts of added ammonium sulfate solution result in different end pH values for the system and potentially different hardening conditions. Drugs for incorporation may become insoluble and precipitate during this pH drift which will lead to nonuseful turbidity and in the extreme situation, microencapsulation.

In weather maps, plots of isobars are made. If the isobars are closely packed together, abrupt changes in weather will occur. If the isobars are well separated, steady uniform weather occurs. Alternatively, contour lines on a map readily indicate cliffs and plateaus. An analogous plot can be made for the desolvation of proteins.

Just as the solubility of proteins is at a minimum at the isoelectric point of the protein and the coacervation concentration is also at a minimum at this pH, the extent of light scattering from variously desolvated proteins is a minimum at this pH. Figure 4 is a schematic representation of the saturation and coacervation concentrations as well as lines of equal light scattering called isonephs. An isoneph line is created by measuring the extent of light scattering with a nephelometer in a number of experiments, in this case at various pH values, and then joining points of equal light scattering intensity to create the isoneph.

Figures 5 to 10 are combinations of Figures 1 and 2 in that isonephs have been created from a series of light scattering desolvation profiles recorded as different starting solutions of BSA have been desolvated. The composition of the partially desolvated system which gives the predetermined readings of 6, 10, and 30 on the Turner nephelometer has been converted into points on a three-component diagram. The isoneph plot was made by joining the proportion points giving equal light scattering.

Figures 5 to 7 represent drug-free systems at 37°C with increasing amounts of Polysorbate 20. The systems containing no Polysorbate 20 shown in Figure 5 have well-separated isonephs, indicating only a gradual change in the system as the desolvation process is carried out. For systems initially containing around 6% BSA, a ridge appears on the isoneph map. If the desolvation line ran along this ridge, there would be little change in the light scattering as the desolvation proceeds. Hence, the choice of starting proportion of BSA will determine whether the light scattering-desolvation profile initially goes down, remains constant, or slowly drifts upward.

The addition of small amounts of Polysorbate 20, corresponding to the amounts most commonly used in the preparation of nanoparticles and represented in Figure 6, leads to an abrupt change in the isoneph pattern, especially at very low proportions of BSA. The ridge has disappeared and the isonephs now seem to represent the edge of a hill. The BSA appears to be more readily desolvated in the presence of the Polysorbate 20. However with further additions of Polysorbate 20, as shown in Figure 7, the isonephs at the higher proportions of BSA are now more compact while there is little change in the isonephs corresponding to the lower proportions of BSA.

It is tempting to maximize the yield of nanoparticles by using a starting system with an increased ratio of BSA to Polysorbate 20. As these Figures show, this ratio has a significant influence on the general shape and slope of the desolvation profile as well as the way the ideal point for hardening is approached or at least defined in terms of the coacervate phase boundary. As we increase the proportion of BSA in the systems, the desolvation process is more easily controlled as compared to the lower BSA proportions. At these lower proportions, a small addition of ammonium sulfate solution gives a dramatic change in nephelos number as the coacervate region is approached. At higher proportions of BSA, the isonephs are not so closely packed. When making nanoparticles, tightly packed isonephs increase the difficulty in getting a reproducible extent of desolvation. Hence, it is comparatively easy to add too much desolvating agent.

Figures 8 to 10 represent the same starting systems as in Figures 5 to 7 but with 0.4% sodium salicylate added. The same general patterns of the effect of the proportion of Polysorbate 20 are seen in these more complex systems. It would also seem that the addition of sodium salicylate has made the BSA more able to be desolvated.

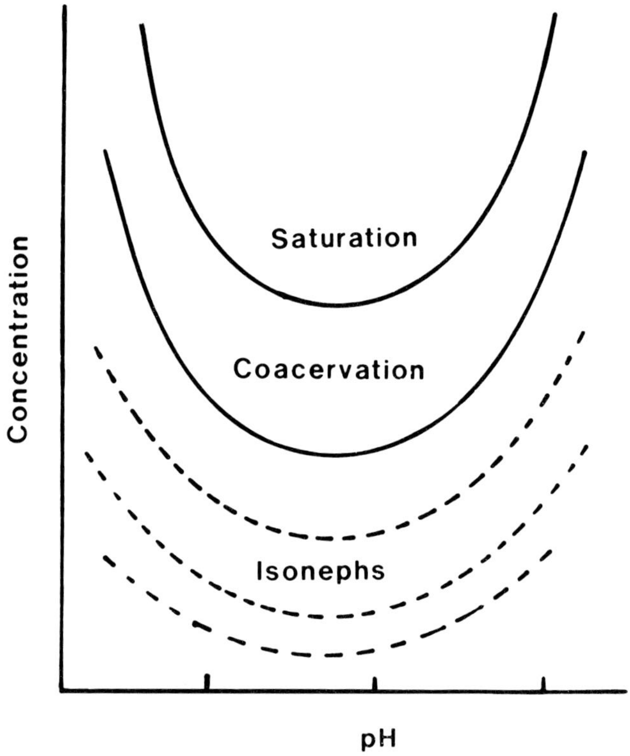

FIGURE 4. Schematic representation of effect of pH on saturation and coacervation concentrations and isonephs for a protein.

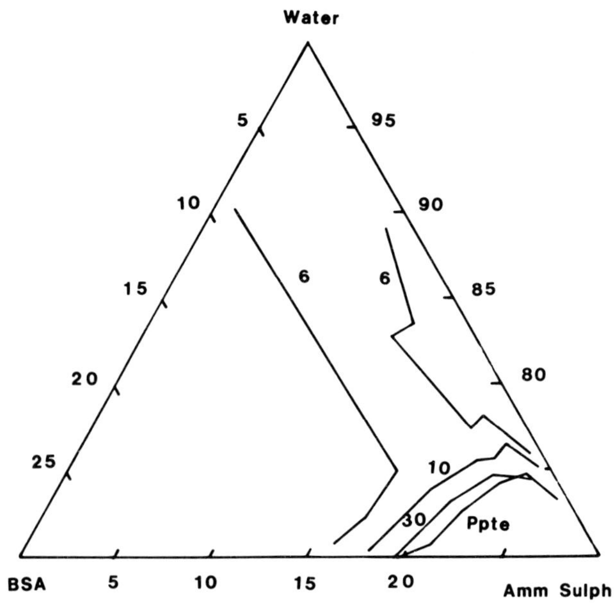

FIGURE 5. Three-component isoneph diagram for aqueous solutions of bovine serum albumin being desolvated with a 40% ammonium sulfate solution at pH 7.0.

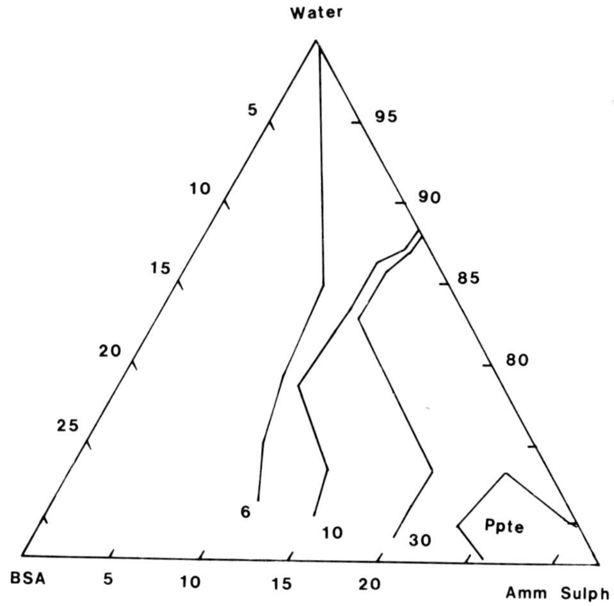

FIGURE 6. Three-component isoneph diagram for aqueous solutions of bovine serum albumin and 2% Polysorbate 20 being desolvated with a 40% ammonium sulfate solution at pH 7.0.

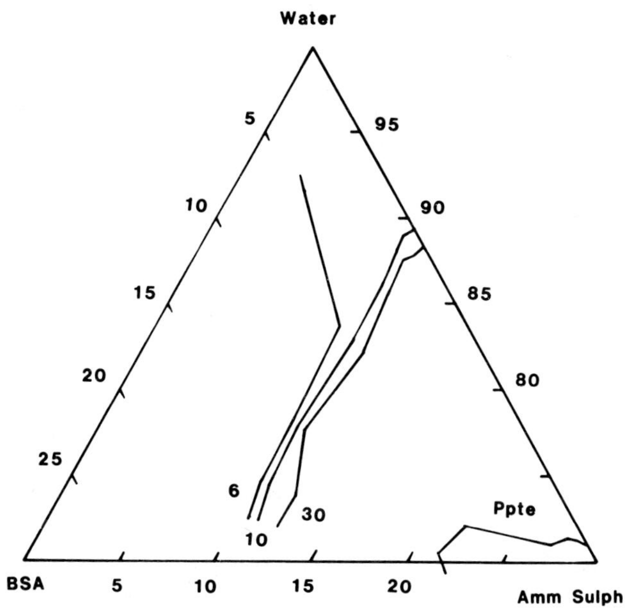

FIGURE 7. Three-component isoneph diagram for aqueous solutions of bovine serum albumin and 4% Polysorbate 20 being desolvated with a 40% ammonium sulfate solution at pH 7.0.

The addition of either or both of Polysorbate 20 or sodium salicylate to BSA solutions seems to block off hydrophilic sites on the macromolecule and the preferred state of desolvation is achieved with the addition of less desolvating agent than needed for a simple BSA solution.

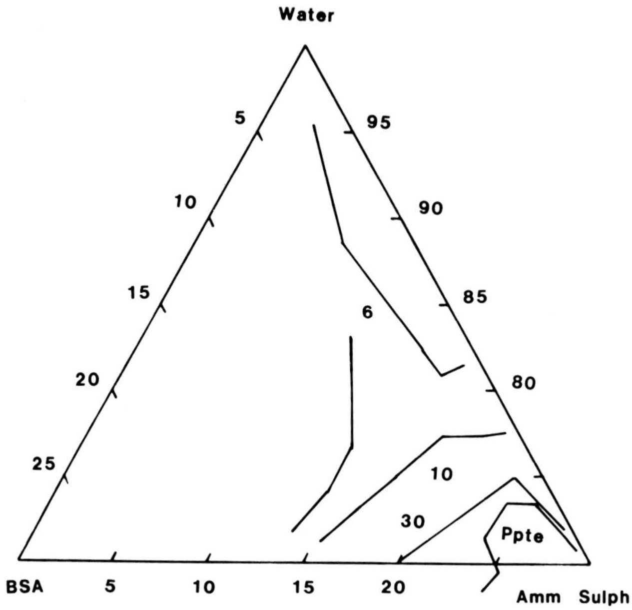

FIGURE 8. Three-component isoneph diagram for aqueous solutions of bovine serum albumin and 0.4% sodium salicylate being desolvated with a 40% ammonium sulfate solution at pH 7.0.

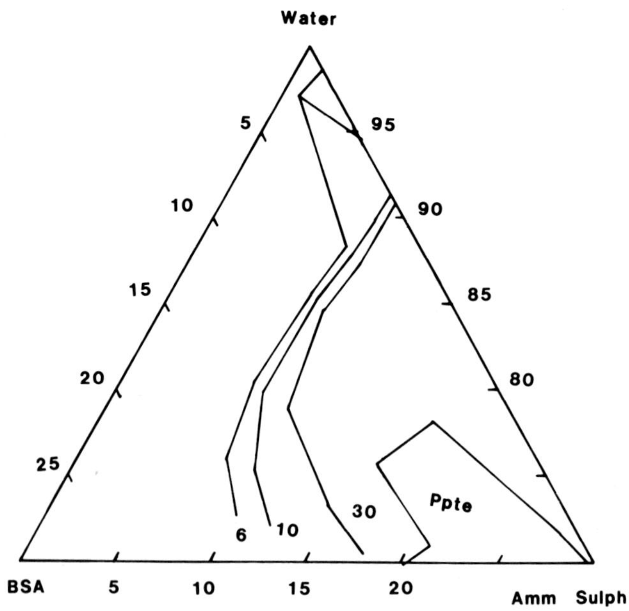

FIGURE 9. Three-component isoneph diagram for aqueous solutions of bovine serum albumin, 2% Polysorbate 20, and 0.4% sodium salicylate being desolvated with a 40% ammonium sulfate solution at pH 7.0.

The use of isoneph patterns appears to be a novel way of studying the complex behavior of macromolecules used in mixtures. The isoneph diagrams can be used to define relative proportions of macromolecule, added surfactant, and water in the start-

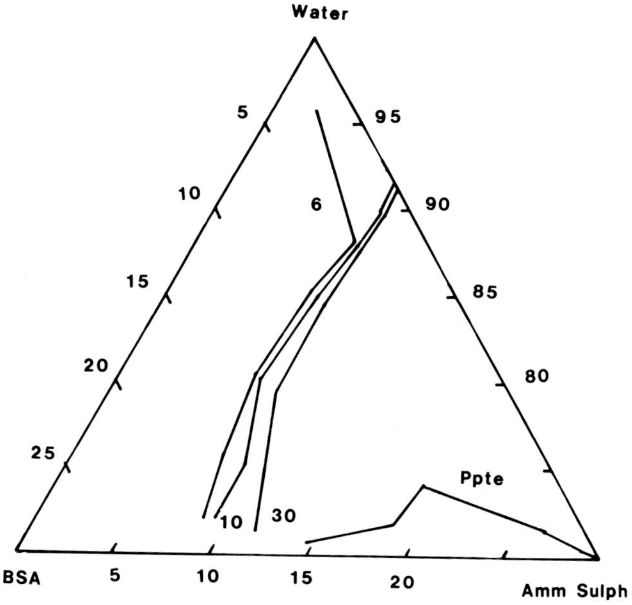

FIGURE 10. Three-component isoneph diagram for aqueous solutions of bovine serum albumin, 4% Polysorbate 20, and 0.4% sodium salicylate being desolvated with a 40% ammonium sulfate solution at pH 7.0.

ing systems which will enable an easy approach to and definition of a suitable extent of desolvation of the macromolecule. With this extent of desolvation readily obtained, as sensed by the intensity of light scattering, nanoparticles may be made to use as a drug delivery system.

IV. GENERAL METHOD OF MAKING PROTEINACEOUS NANOPARTICLES

As was indicated in the preceding section, the amount of desolvating agent needed is dependent upon the exact proportions of the various reagents used. Similarly, the exact amounts of the other reagents need to be defined for any particular system being developed. As a general rule, the amount of glutaraldehyde given below is the maximum routinely used and, as can be seen from other examples in this chapter, the amount is often less than this maximum.

The type of gelatin used does not alter the general procedure but routinely we use Sigma Type II. We also use Fraction V of the albumins. All reagents are analytical grade or pharmacopoeial standard, whichever is the more stringent specification. El-Samaligy and Rohdewald have found[21] that a pH range of 5.5 to 6.5 is optimum to produce uniform-sized gelatin nanoparticles.

A solution (10 mℓ) of 1% gelatin and 0.5% Polysorbate 20 was made and equilibrated at 35°C. Seven mℓ of 20% w/v sodium sulfate were added while stirring with a magnetic stirring bead until the intensity of the scattered light, as monitored by a Turner Designs Nephelometer, rose rapidly and the system acquired a faint permanent turbidity. Isopropanol (1.2 mℓ) was added until the turbidity disappeared and the intensity of the scattered light decreased to the predetermined level. A Silverson Laboratory Mixer Emulsifier, fitted with a 0.625-in. head was used to agitate the system as 0.6 mℓof a 25% aqueous solution of glutaraldehyde were added in 1 aliquot. The hardening process was allowed to continue for 20 min at 35°C when it was terminated by the addition of 5 mℓ of a 12% sodium metabisulfite solution. The crude mixture was shell frozen in a dry ice/acetone bath and lyophilized overnight on a FD_2 Dynavac Freeze Drier. Upon reconstitution in 10 mℓ of 0.04% chlorbutol solution, the mixture was desalted on a Sephadex G-50 m column. The void volume containing the nanoparticles may then be freeze dried and either stored ready for testing or reconstituted and further fractionated on another size exclusion column such as Bio-Gel A5m. The Sephadex clean up is all that is routinely done in our laboratory. Scale up of the purification procedures to industrial size batches is not expected to present any insurmountable problems.

Once the nanoparticles are sufficiently purified they can be evaluated in a number of ways.

Scanning electron microscopy — The criterion of success is the uniformity of size of submicron particles. Excessive crosslinking leading to chains or lumps of aggregated particles is symptomatic of inappropriate desolvation either to the wrong extent or at the wrong pH,[21] excessive amounts of glutaraldehyde added, or insufficient killing of the hardening reaction. Despite increasing the magnification to at least 20,000 ×, no visual evidence of porosity has ever been seen for the nanoparticles made by this method. This is in direct contrast to many of the other solid colloidal drug-containing particles discussed elsewhere in this book.

Redispersion — In order that the maximum amount of drug be administered in the minimum volume of injectable liquid, the freeze-dried nanoparticles must be reconstituted with as little liquid as possible. A 10% dispersion is considered to be the lower limit of extent of useful dispersion. A 25% dispersion is about the maximum practical extent that can be achieved. The economic and therapeutic limitations on the extent of dispersion are dependent upon the free drug solubility and the relative potency of the drug either as a free solution or as a nanoparticle dispersion.

V. PAYLOAD CONSIDERATIONS

When more drug can be incorporated into each unit weight of delivery system, then less of the delivery system is required for administration and targeting of a given dose. The maximum payload that can be achieved in nanoparticles is close to 100%. This would occur if the macromolecule forming the nanoparticle had intrinsic biological activity. Insulin, discussed later, is an example of such a macromolecule.

However, most proteinaceous nanoparticles are made up of a carrier macromolecule such as gelatin or albumin and the incorporated drug. In this situation if the payload approaches 30% on a weight basis, it is hard to conceive how the drug can be kept away from the external environment. It is considered that with such a high payload, the drug would dissolve forming channels and pores into the nanoparticle. Hence, the release profile would be at least biphasic and the portion being released at the slower rate would represent the effective payload most likely to be delivered to the target site. Couvreur et al.[22] discuss this exchangeable payload for porous polyalkylcyanoacrylate nanoparticles containing dactinomycin. It is hard to predict what figure would represent the maximum payload achievable in nonporous proteinaceous nanoparticles above which exchange may become a problem. In the systems produced in our laboratory, we find that we can achieve up to a 15% payload without evidence of any exchange occurring in developing pores. While this represents an upper limit on payloads achieved, many drugs cannot be incorporated to nearly such a high level. There are at least five factors which could influence the payload of proteinaceous nanoparticles.

Table 1
PAYLOADS OF BSA NANOPARTICLES CONTAINING ASPIRIN WHEN DIFFERENT AMOUNTS OF ASPIRIN WERE ADDED TO 500 mg OF BSA AND 200 mg POLYSORBATE 20 AND NANOPARTICLES MADE AT pH 7.0

Aspirin (mg) in starting solution	% Payload achieved
50	0.6
100	3.0
200	7.3
300	11.3
400	16.2

A. Extent of Protein Binding

If the macromolecule is albumin and the drug to be incorporated is highly protein bound, then there is a good chance of a high payload. If there is poor interaction between the drug and the macromolecule, then although satisfactory nanoparticles might be made, the payloads will be disappointingly low. Drugs do not bind as strongly to gelatin as they do to albumin and so the payloads will be lower for gelatin-based nanoparticles.

An increase of the drug to protein ratio will increase the payload. Table 1 gives the payloads achieved when increasing amounts of aspirin were added to a fixed amount of BSA and then nanoparticles made. While the efficiency of the incorporation process may decrease (i.e., the percentage of the initially added drug that ends up as useful payload decreases) this is not as important as the increase in payload. Provided this loss of efficiency is not a cost limiting factor in the overall product manufacture, product administration and health care of the patient, an increase in drug to protein ratio is likely to lead to a more useful product.

If the drug binds irreversibly to the carrier protein, then although adequate nanoparticles with an acceptable apparent payload might be able to be made, such a delivery system would not be very efficacious. Drugs containing reactive groups such as aldehyde are likely to be a problem in this regard.

Protein binding of drugs is known to be pH dependent. Adjustment of pH in a nanoparticle reaction system can occur for a variety of reasons including increasing the drug solubility, controlling the hardening reaction, or trying to minimize drug degradation. Appropriate compromises may have to be made between extent of binding and such other reasons. The effect of desolvation upon protein binding is discussed under the fourth factor affecting payloads.

B. Drug Solubility

Usually the first attempt to make a proteinaceous nanoparticle containing a drug is to desolvate the protein from an aqueous environment. If the drug to be incorporated into the nanoparticle has a low solubility at the pH at which the system is initially equilibrated, then the payload in the nanoparticles is likely to be low. There is the additional risk of microencapsulating the undissolved drug crystals. To increase the number of drug molecules that are free to interact with the macromolecule three types of action may be taken.

First, if the drug is an acid, its solubility is increased at high pH. However, to achieve a sufficiently high solubility to ensure an acceptable payload in the final nanoparticles, it may be necessary to increase the pH to 8.5 or beyond. At these higher pH regions, the hardening reaction becomes very rapid and difficult to control. Hence, the pH must be reduced to below 8.0 before hardening is commenced. A flukicide F300 is such an acid drug. Aqueous solutions of 0.1% of the drug could only be formed at pH values in excess of 11.0. To incorporate this drug the following formula and procedure was used:[9]

Human serum albumin	500 mg
Polysorbate 20	200 mg
F 300	10 mg
Water	to 10 mℓ

The system was equilibrated at 37°C and at pH 9.0. A clear solution was rapidly formed. Upon adding dilute hydrochloric acid dropwise the pH was reduced to 7.5 and the system remained clear. The protein was desolvated with 7.0 mℓ of 20% sodium sulfate solution and resolvated with 15 drops of isopropanol to form a clear solution ready for hardening. This was achieved by reacting with 0.2 mℓ of 2.5% glutaraldehyde for 5 min before adding excess sodium metabisulfite to terminate the hardening. At no stage did the drug precipitate nor did it appear to be associated with the Polysorbate 20 in the purification step.

The second way of increasing the number of drug molecules free to interact with the protein is to modify the solvent. Many of the large molecules used in modern therapeutics are more soluble in aqueous ethanol or aqueous glycerol than they are in pure water. There appears to be no significant hindrance to making nanoparticles in these mixed solvent systems. However, the amount of electrolyte type desolvating agent may be drastically reduced or cannot be used at all because it becomes a saturated solution before the protein is sufficiently desolvated. In this case desolvation can be attempted with alcohols. It is possible to add the drug as an alcoholic solution so making the initial co-solvent vehicle and then continuing the desolvation of the protein with ethanol. In this way the drug remains in a nonprecipitated form but the protein is increasingly desolvated. A 5-mg lot of triamcinolone acetonide was dissolved in 1 mℓ of ethanol. This drug solution was then slowly added to a solution of 500 mg BSA and 500 mg of Polysorbate 20 in 9 mℓ of water. The resultant clear solution was successfully desolvated with a further 1.2 mℓ of pure ethanol and then the system hardened with 0.2 mℓ of 5% glutaraldehyde for 10 min before terminating the hardening with a sodium metabisulfite solution.[23]

While pH measurements in mixed solvent systems are difficult to interpret, there should be no practical reason why pH changes could not be used in conjunction with solvent changes to maximize payloads.

The third major way drug solubility can be modified to increase payloads is to use a surfactant in combination with the macromolecule. The solubility of drugs increases markedly when the surfactant concentration exceeds the critical micelle concentration. With the drug present as discrete molecules, interaction with the protein can be more rapid and more efficient. Provided the drug does interact with the macromolecule and doesn't simply remain solubilized in the surfactant micelle, then the payload of the resultant nanoparticle will be increased.

For maximum flexibility of the process, a nonionic surfactant would be better than a charged surfactant. Of the many such surfactants available, our group has used the Polysorbate series. These surfactants were chosen because of a very low parenteral toxicity, acceptance for use in established parenteral products by regulatory authorities, and their ability to solubilize a wide range of drugs. Initially, Polysorbate 80 was added but its cloud point was reduced to below the working temperature of 35°C upon the addition of desolvating agents, causing turbidity before the protein was sufficiently desolvated. The shorter hydrocarbon chain of laurate used in Polysorbate 20 means the molecule is more hydrophilic and the cloud point is well above the working temperature. No premature desolvation problems have been encountered.

The major limitation to large amounts of surfactant being used to increase the solubility of insoluble drugs considerably seems to be the problem of frothing which occurs upon stirring during the hardening stages of the nanoparticle production.

Using the maxim that any procedure which will increase the drug solubility is likely to increase payload, combinations of pH changes, solvent changes, and surfactant type and proportion should be investigated for any particular system under study.

C. Drug Stability

The aim of incorporating a drug into nanoparticles is to deliver that drug to its desired site of action. If the drug degrades during the manufacture procedure then there is the risk of incorporation and delivery of the wrong molecule. It is recognized that binding the drug to a macromolecule can enhance its stability.[24] While pH is the obvious parameter to control, care needs to be exercised if nucleophilic attack of the drug by ethanol for example is a possibility. Many oxidative degradations are catalyzed by light, in particular ultraviolet (UV) radiation. However since the final product contains the drug molecularly dispersed throughout the solid nanoparticle, it is unlikely that light will cause degradation of the final finished product.

There are three basic ways of converting individual carrier molecules into particulate aggregates. The irradiation method used in the synthetic polymer-based nanoparticles is not relevant to proteinaceous nanoparticles. However the chemical and thermal methods can lead to a loss of potency and so a reduced payload. Glutaraldehyde is most commonly used as the chemical crosslinking agent for the protein but it can also react with amine groups on drugs. Methotrexate cannot readily be incorporated because it reacts with glutaraldehyde. However, adriamycin, which also contains an amine group, does not react with glutaraldehyde added to the desolvated protein, drug and Polysorbate 20 mixture. It is not clear why one cytotoxic drug can and one cannot be incorporated but it must have something to do with the accessibility of the amine group to the glutaraldehyde.

The heat-aggregated protein aggregates may contain heat degraded drug. The working temperature in our process is 35 to 37°C and the drugs are only exposed to this temperature for about 1 hr. No problems of this type have been identified in any of the drugs used in our program.

D. Desolvation

The desolvation process may lead to a change in payload in two ways. It is probable that the change in solvent upon the addition of, say, ethanol as the desolvating agent, will alter the solubility of the unincorporated drug. If this solubility is reduced, it is also possible that the amount of bound drug in equilibrium with the free drug will be changed leading to an unexpected payload.

The other way desolvation may influence payload is to change the parts of the macromolecule which are exposed to the vehicle and hence change the pattern of protein binding. Despite the mass of data reported on protein binding, there appears to be little on the effect of desolvation on protein binding.

Using sodium salicylate as a model drug, we used ultrafiltration at 22°C from sacks of seamless cellulose dialysis membrane to make some preliminary investigations on the effect of desolvation on binding the drug to BSA and Polysorbate 20 solutions.

Sodium salicylate did not bind to the membrane. A nonlinear regression computer program was used to fit the data to a generalized Scatchard type binding expression.

Using BSA and Polysorbate 20 concentrations typical of nanoparticle production and sodium salicylate concentrations from 0.2 to 75 mg/mℓ, the data best fit a two- rather than a three-binding site model. Although experimental difficulties prevented us from doing the ultrafiltration at 35°C, we are hopeful that the results at these high drug concentrations are indicative of binding at the higher temperature. Wosilait[25] found that temperature had little effect on the binding of salicylate to BSA provided the total drug concentration exceeded 1 mg/mℓ.

We found that desolvation with ammonium sulfate solution did not significantly alter the number or strength of the primary binding sites. However, the number and the strength of the secondary binding sites were dependent upon the extent of desolvation and the changes were influenced by the amount of Polysorbate 20 present. Approximately 4 mℓ of ammonium sulfate solution were required to desolvate the system sufficiently for nanoparticle production. The addition of 1 mℓ of this solution caused a two- or threefold drop in the number of secondary binding sites but a fourfold increase in their association constant. At the normal 2% level of Polysorbate 20, no further changes occurred when up to 3.5 mℓ of ammonium sulfate solution were added. However, at the very high level of 10% Polysorbate 20 when 3.5 mℓ of ammonium sulfate solution were added, the number of secondary binding sites was not significantly different to the nondesolvated situation but the association constant was slightly higher than initially.

It is suggested that the desolvation process is causing some change in the conformation of the protein in solution and there is a loss of binding sites in the early stages of the desolvation. The addition of Polysorbate 20 to the system may lead to competition between the Polysorbate 20 and the sodium salicylate for binding sites on the BSA. This would cause a decrease in payload with increasing Polysorbate 20. Alternatively, the Polysorbate 20 may bind to BSA at sites that do not bind the sodium salicylate and induce structural changes in the tertiary conformation of the protein.[26] Such changes can alter the number or the affinity of accessible binding sites available to the drug.

Further detailed studies need to be made to define the effect of desolvation upon the binding of a wide range of drugs to proteins in the presence of surfactants so that predictions about the consequences for payload can be accurately made.

E. Hardening

The reaction of glutaraldehyde with amine groups causes the desolvated protein to be fixed and nanoparticles to form. Mention has already been made of unwanted glutaraldehyde-drug reactions which will lead to a reduced payload. There are two other ways that the useful payload can be influenced by the hardening step.

It is common practice to add the glutaraldehyde to the proteinaceous system and allow the reaction to proceed. Manufacturing folklore has it that if more than 4% glutaraldehyde is added, then the particles, be they micro- or nanometer in size, are not readily biodegraded. This is attributed to excess crosslinking in the particles. These poorly degraded particles can deliver their payload of drug to the target site but the drug is not released. Hence the effective payload is very low. The way to overcome this problem is to only allow the hardening reaction to proceed for a limited time and then to react the excess glutaraldehyde with sodium metabisulfite solution. The resultant addition compound is eluted with the low molecular weight species in the size fractionation chromatography clean up of the product.

The second way the hardening reaction can influence the payload is by altering the surface characteristics of the nanoparticle. Harrington and Barry have found[27] that hardening for more than about 4 min causes an increase in the hydrophobic character of the surface. This very interesting in vitro result could mean that in vivo there will be a variety of opsonization processes on the surface of the particle and that could lead to uptake by a variety of cells. Little appears to be known of the targeting effect of opsonization of particles in general and more specifically of proteinaceous nanoparticles.

The payload of all the systems developed so far has been determined indirectly. The purified nanoparticles are obtained from the void volume of the gel permeation column. Later fractions contain the unincorporated drug. Knowing the amount of drug initially put into the reaction mixture, the amount in the nanoparticles is obtained by difference. The payload figure is then calculated as the number of milligrams of drug contained in 100 mg of final delivery system. Allowance for the losses of drug by means of degradation and binding to the equipment and Sephadex have to be made. The payload calculated represents an upper limit. If for some reason it is less than that calculated, this means that the potency of the delivery system is higher than that derived from the in vivo testing.

Drugs which form large molecular weight aggregates will cause unincorporated drug to be eluted in both the void and the salt peaks. Further work needs to be done to separate and quantify the amount in the void volume so that the payload of the nanoparticles can be calculated and there is no chance of unincorporated drug being administered to dilute the targeting capacity of the nanoparticles.

It would be preferable to use radioactive drugs so that the payload could be calculated directly. Alternatively, a nanoparticle sample could be enzymatically degraded in vitro and the amount of drug released from a defined weight of nanoparticles calculated. Our group has deferred such refinements until a delivery system appears to be intended for large scale clinical trialing. El-Samaligy and Rohdewald[21] have shown that gelatin nanoparticles can be degraded by trypsin, collaginase, and trypsin/collaginase mixtures.

VI. SPECIFIC APPLICATIONS OF PROTEINACEOUS NANOPARTICLES

This chapter has so far concentrated on obtaining a high payload of drug by having the drug molecularly dispersed throughout the nanoparticle. Before considering applications of these systems, attachment of molecules to the surface of the nanoparticle is reviewed.

The payload that can be achieved by surface conjugation must be lower than internal incorporation of the drug. However, definition of the surface sites available for conjugation is important if directing devices such as antibodies are to be attached to the surface.

Proteinaceous nanoparticles can be surface labeled with ^{99m}technetium.[28] Using a tin (II) reduction procedure, the labeling can be up to 94% efficient. The time of reaction did not appear to be so critical, but since the yield of labeled particles was slightly higher after a 15-min reaction period, this time was used for all later work. The other significant factor was the importance of adjusting the final pH very carefully. If the 6.0 to 6.5 region is overshot, then the labeling efficiency is as low as 35%. When injected intravenously into rats, there is a rapid accumulation in the liver as would be expected of a colloidal system. Following intramuscular and intraperitoneal administration, about 5 and 3% of the radioactivity was found in the liver respectively 30 min after administration.[14] However, if the tagged nanoparticles are injected into the knee joints of non-arthritic rabbits,[29] 90% of the radioactivity (corrected for decay) is found in the joint 24 hr after administration. There was a small accumulation of radioactivity in the thyroid 1 hr after administration and this was attributed to unbound technetium which could rapidly diffuse out of the joint. No other organ showed any accumulation other than the background due to the small amount of radioactivity in the blood. In principle, this technetium tagged system may be used to study in vivo distribution following any route of administration.

Fluorescein isothiocyanate (FITC) is well known to conjugate to the ε-amino group of lysine. Protein conjugates are widely used in clinical pharmacology. We found[30] that FITC could be conjugated to the surface of gelatin and albumin nanoparticles. A variety of tumor cells including EMT-6, WEHI-3, SP-1, B-16 and a virus-induced mouse mammary tumor have all been separately incubated[30,31] with FITC conjugated nanoparticles. Clear nuclear shadows have been seen indicating that the nanoparticles have been taken into the cells rather than being adsorbed on their surface.

The technetium labeling experiments show that there are surface hydroxyls available for drug binding. Similarly, the FITC conjugation is to amino groups on the surface. These two types of surface groups need to be exploited to chemically conjugate a wide range of molecules and the products tested in vivo. El-Samaligy and Rohdewald[21] have shown that cytotoxics and antiarthritic drugs can be reversibly adsorbed onto the surface of gelatin nanoparticles. Modifications of the surface character along the lines outlined by Harrington and Barry[27] may enable different drug types to be adsorbed to different extents. Control of the desorption process may even be possible.

If nanoparticles can be made out of biologically active macromolecules and the manufacturing process does not destroy the activity, then payloads approaching 100% are possible. Following all the initial excitement and then disappointment of orally administered insulin containing liposomes in the late 1970s, we attempted to make nanoparticles based on insulin alone arguing that solid particles would be more likely to survive in the gut than fluid liposomes.[13] Batches of insulin nanoparticles were made from 10-mℓ vials of Neutral Insulin Injection (Actrapid, 80 U/mℓ, Novo). After 10 mℓ of the solution were equilibrated at 25°C, about 0.7 mℓ of 0.1 *M* hydrochloric acid was needed to cause a rapid rise in light scattering. Resolvation was done by adding 0.1 mℓ of 0.1 *M* sodium hydroxide solution; 0.2 mℓ of 25% glutaraldehyde solution was added in 1 aliquot and the system stirred vigorously for 15 sec every 30 sec for a total of 4 min after which 6 mℓ of 12% sodium metabisulfite was added. The same intermittent stirring pattern was used for a further 6 min.

After thorough cleanup, the 200 nm insulin nanoparticles were tested in vivo. Since the nanoparticles would have to be broken down before any significant antiglycemic activity could be shown, the nanoparticle form would be expected to have a slower onset of action than the free form. Intravenous administration of the nanoparticles showed that a large proportion on the biological activity of insulin was available from the nanoparticular form but its onset was indeed slower than the original Actrapid insulin.

The nanoparticles were then administered to the intestinal tract of mice and of normal and diabetic rats. The blood glucose concentrations in some animals could be reduced to about 15 to 20% of the starting level, 3 hr after administration of between 35 and 70 mg of nanoparticles per 100 g body weight. This was more pronounced than similarly administered Actrapid. However, the dose of insulin needed in the nanoparticle form was too high to warrant further industrial development.

This study, however, does show that biological activity can be retained when macromolecules are converted to nanoparticles. With the current surge in genetic engineering, there are likely to be biologically active macromolecules which may be better delivered in a particulate rather than a free form.

As nanoparticles, along with all other colloidal drug delivery systems, will be taken up by phagocytic cells in the liver, parasitic diseases of the liver could well be treated by nanoparticles. The immature form of *Fasciola hepatica* does considerable damage to livers by burrowing. If it could be killed before causing too much damage, there would be considerable economic advantages in the sheep and cattle industry. There would also be many reasons to eradicate this infestation in humans but regrettably such a health care program is economically unattractive. Most flukicides have a low therapeutic index and recent developments have concentrated on potent decouplers of oxidative phosphorylation.

Nanoparticles, based on human serum albumin, containing an experimental flukicide 10645 were made and tested against an immature form of the fluke in rats.[32] The drug is a halogenated trinuclear phenol and has a low solubility in water at pH 7. Nanoparticles were made according to:

Human serum albumin	5.0 g
Polysorbate 20	2.0 g
10645	100 mg
Water	to 100 mℓ

This system was equilibrated at 35°C and to the resultant clear solution 57.3 mℓ of 20% sodium sulfate solution were added slowly and dropwise to desolvate the protein. If the addition was done quickly the free drug appeared to gel. Hence, time was needed for the equilibrium between the drug and its binding sites on the desolvating protein to be established. Fifty-three drops of water resolvated the system and then it was divided into five portions each of which was hardened for 5 min with 0.2 mℓ of 2.5% glutaraldehyde solution added in 1 aliquot. After termination and clean up, the 10645 nanoparticles had a payload of about 4.5%. When dispersed in 3 mℓ of normal saline this gave an effective solubility for the drug of about 1.5% which is a least two orders of magnitude greater than for the pure drug.

When compared to a complex solution of the drug at pH 10, both products exhibited about equal efficacy following either subcutaneous or intravenous administration. The nanoparticle product could be used at twice the dosage of the control formulation before toxic reactions appeared. Hence, incorporation into nanoparticles has not impaired its efficacy and has doubled its therapeutic index.

Despite this work no other groups appeared to have attempted to use nanoparticles to treat immature liver fluke.

New et al.[33] have shown that antimonials can be entrapped in liposomes and used to treat leishmaniasis, a parasitic disease of macrophages. BSA nanoparticles can be made containing sodium stibogluconate according to:

BSA	500 mg
Polysorbate 20	200 mg
Sodium stibogluconate	100 mg
Water	to 10 mℓ

Desolvate with 4.0 mℓ of 40% ammonium sulfate solution at 37°C and then resolvate with 0.2 mℓ of isopropanol. The 10-min hardening reaction is done with 0.2 mℓ of 5% glutaraldehyde. This was terminated with 1 mℓ of 12% sodium metabisulfite. A payload of 9.1% Pentostam (equivalent to 2.9% Sb(v)) was achieved.

A related product was made with pentamidine isethionate using a mixed solvent.

BSA	500 mg
Polysorbate 20	200 mg
Pentamidine isethionate	100 mg
Glycerol/water (1/3) at pH 7	to 10 mℓ

Desolvation was done with 2.0 mℓ of 95% alcohol at 37°C. No resolvation was possible because the system became irreversibly gelled if excessive desolvation was done with the alcohol. The hardening and termination reactions were as for the Pentostam. A payload of 3.9% was achieved.

Neither product showed any significant efficacy in a preliminary screen against a subcutaneously implanted infection in rats. Further in vivo testing of this type of product is currently underway.

If a joint is inflamed, the most efficient way of treating it is to give an intra-articular injection of a suitable drug. Having got the drug into place, it is imperative to keep it within the synovial cavity. Blood levels of injected steroids can be detected very soon after injection and reach a peak within hours.[34] If an antiarthritic drug could be incorporated into nanoparticles then it is not likely to be removed from the joint and the delivery system is likely to be taken up by the invading phagocytic cells, degraded, and the drug released and so prevent the release of the irritating enzymes. The triamcinolone acetonide product described earlier only had a payload of less than 1% and as such did not reduce joint swelling induced in rabbit knees by carrageenan.[23] A program has been initiated to incorporate gold salts and other antiarthritic drugs into gelatin based nanoparticles in an attempt to develop an effective, nonimmunogenic intra-articular product.

Since the FITC conjugated nanoparticles can be taken up by a variety of tumor cells, a series of cytotoxic-containing nanoparticles were prepared and tested in vivo.[31] 5-Fluorouracil and doxorubicin were separately incorporated into human serum albumin nanoparticles. The payload for the 5-fluorouracil nanoparticles was around 8% which compares favorably with the 0.25% found by Hashida et al.[35] for their emulsion type system and the 2.3% payload found by Morimoto et al.[36] for their heat-stabilized albumin microspheres. After making the doxorubicin containing nanoparticles, the salt peak was first cleaned up by high pressure liquid chromatography using a C18 reverse-phase column before the amount of free unincorporated drug was assayed by fluorescence spectroscopy. An unoptimized payload of 2.5% was achieved. This compares to that of 4% found by Widder et al.[37] but lower than the 14% reported by Couvreur et al.[38] for their methylcyanoacrylate nanoparticles.

In vitro dose response and growth kinetics testing of both products against B16 melanoma and the virus-induced mouse mammary tumor, MMTV showed that the nanoparticle product could exert both a cytotoxic and a cytostatic effect depending upon concentration. Compared to free drug, the in vitro potency varied between 50 and 10%. This may be due to less of the drug reaching the interior of the cell because of slow diffusion in the medium of the nanoparticle product or incomplete degradation of the nanoparticle. A full discussion is available elsewhere.[39]

In vivo, both dose response and acute toxicity studies have been done. The therapeutic index of free 5-fluorouracil against B16 melanoma implanted subcutaneously in mice was around 1.5. That of the nanoparticle product was about 2.5. Similarly, the therapeutic index for the doxorubicin system rose from 1.6 to around 3.0. The nanoparticle delivery of doxorubicin would also be likely to decrease the cardiotoxicity of the drug which is one of its major drawbacks upon chronic administration. Again a full discussion is available elsewhere.[39]

It is believed that these examples of the applications of proteinaceous nanoparticles illustrated the versatility of this type of drug delivery system.

VII. PROBLEMS TO OVERCOME

As soon as a protein-based delivery system is proposed, it has to be recognized that there may be immunological complications. Macroaggregated albumin has been used for many years as a diagnostic agent and more recently as a drug delivery system with insignificant incidence of adverse reactions. Initial work with drug-free gelatin and bovine serum albumin nanoparticles injected repeatedly intravenously into dogs and rodents could show no gross antigenic response.[19] Subsequent work with cytotoxics in human serum albumin-based nanoparticles injected by the intravenous, intraperitoneal, subcutaneous, and intramuscular routes likewise showed no adverse response.[28] However, it is possible that the cytotoxic may be masking an immune response. Intra-articular injection of antiarthritic drugs in BSA nanoparticles did cause increased joint diameter[23] which was not unexpected since BSA macroaggregates are sometimes used to induce experimental arthritis. However, the same drugs in gelatin nanoparticles did not cause any changes in the diameter of normal joints. No adverse immunological response was seen with the oral administration of insulin based nanoparticles.[13]

While these results are encouraging, more work is required to define the type of macromolecule and the routes of administration that will give no immunological problems. Presumably simple immunoprecipitation would be the first step in such a program. This has not been done pending the development of a delivery system containing a useful payload of drug which looks as if it may be worthwhile taking to a large scale clinical trial.

The size of the nanoparticles may also be a critical factor. Illum[40] believes that only particles less than 100 nm are likely to be able to leave the systemic circulation and reach extra-vascular sites such as tumor cells. To investigate this, a series of proteinaceous nanoparticles of different particle size needs to be made, suitably tagged and injected intravenously into tumor-bearing animals and the in vivo distribution evaluated.

This last point is an example of a much broader problem with submicron particles. The reticuloendothelial cells in the liver and the spleen as well as the circulating monocytes have the function of removing such particles from the circulation. If the nanoparticles are to deliver their payload of drug to the desired site of action and not necessarily to these scavenger cells, then reliance upon phagocytic function may be not enough. Work is starting on extra directive procedures such as antibody attachment and inclusion of a payload of magnetic particles inside the nanoparticles to attempt to improve the targeting capacity. Appropriate delivery is important. Intravenous, intramuscular or subcutaneous, while common parenteral routes, may not be the most appropriate for all nanoparticle therapy. All possible routes, compatible with the disease state being treated, need to be investigated.

ACKNOWLEDGMENTS

Financial support for the nanoparticle program from the Australian Research Grants Scheme (C76/15138), the National Health and Medical Research Council (830092), I.C.I. (Australia), Nicholas Drug Research Consortium, and E. R. Squibb and Sons is gratefully acknowledged. I record my debt to the graduate students and technical staff of the College whose work has enabled the progress recorded here possible.

REFERENCES

1. Oppenheim, R. C., Surfactants and micelles in pharmaceutical formulation, *Aust. J. Pharm. Sci.*, 5, 11, 1976.
2. Trouet, A. and Deprez de Campeneere, D., Daunorubicin DNA and doxorubicin DNA — a review of experimental and clinical data, *Cancer Chemother. Pharmacol.*, 2, 77, 1979.
3. Trouet, A., Increased selectivity of drugs by linking to carriers, *Eur. J. Cancer,* 14, 105, 1978.
4. Ryser, R. J. P. and Shen, W. C., Conjugation of methotrexate to poly (L-lysine) increases drug transport and overcomes drug resistance to cultured cells, *Proc. Natl. Acad. Sci. U.S.A.*, 73, 3867, 1978.
5. Kojima, T., Hashida, M., Muranishi, S., and Sezaki, H., Mitomycin C — dextran conjugate — a novel high molecular weight prodrug of mitomycin C, *J. Pharm. Pharmacol.*, 32, 30, 1980.
6. Varga, J. M., Asato, N., Lande, S., and Lerner, A. B., Melanotropin — daunomycin conjugate shows receptor mediated cytotoxicity in cultured murine melanoma cells, *Nature (London),* 267, 56, 1977.
7. Tsuruo, T. and Fidler, I. J., Differences in drug sensitivity among tumour cells from parenteral tumours, selected variants, and spontaneous metastases, *Cancer Res.*, 41, 3058, 1981.
8. Kimura, I., Ohnoshi, T., Tsubota, T., Sato, Y., Kobayashi, T., and Abe, S., Production of tumour antibody-neocarzinostatin (NCS) conjugate and its biological activities, *Cancer Immunol. Immunother.*, 7, 235, 1980.
9. Boag, C. C., Organ Directive Drug Delivery Systems, M. Pharm. thesis, Victoria Institute of Colleges, Melbourne, 1979.
10. Lee, V. H-L. and Robinson, J. R., Methods to achieve sustained drug delivery — the physical approach: oral and parenteral dosage forms, in *Sustained and Controlled Release Drug Delivery Systems,* Robinson, J. R., Ed., Marcel Dekker, New York, 1978, chap. 3.
11. Chien, Y. W., *Novel Drug Delivery Systems,* Marcel Dekker, New York, 1982, chap. 7.
12. Oppenheim, R. C. and Speiser, P., Über die Stabilität kolloidaler Arzneiformen, *Pharm. Acta Helv.*, 50, 245, 1975.
13. Oppenheim, R. C., Stewart, N. F., Gordon, L., and Patel, H. M., The production and evaluation of orally administered insulin nanoparticles, *Drug Dev. Ind. Pharm.*, 8, 531, 1982.
14. Oppenheim, R. C., Nanoparticles, in *Drug Delivery Systems,* Juliano, R. L., Ed., Oxford University Press, New York, 1980, chap. 5.
15. Couvreur, P., Kante, B., Roland, M., and Speiser, P., Adsorption of antineoplastic drugs to polyalkylcyanoacrylate nanoparticles and their release in calf serum, *J. Pharm. Sci.*, 68, 1521, 1979.
16. Tritton, T. R., Yee, G., and Wingard, L. B., Jr., Immobilized adriamycin: a tool for separating cell surfaces from intracellular mechanisms, *Fed. Proc. Fed. Am. Soc. Exp. Biol.*, 42, 284, 1983.
17. Heide, K. and Schwick, H. G., Salt Fractionation of Immunoglobulins, in *Handbook of Experimental Immunology,* 2nd ed., Weir, D. M., Ed., Blackwell Scientific, Oxford, 1973, chap. 6.
18. D'Onofrio, G. P., Oppenheim, R. C., and Batemen, N. F., Encapsulated microcapsules, *Int. J. Pharm.*, 2, 91, 1979.
19. Oppenheim, R. C., Solid colloidal drug delivery systems: nanoparticles, *Int. J. Pharm.*, 8, 217, 1981.
20. Nixon, J. R., Khalil, S. A. H., and Carless, J. E., Phase relationships in the simple coacervating systems isoelectric gelatin:ethanol:water, *J. Pharm. Pharmacol.*, 18, 409, 1966.
21. El-Samaligy, M. S. and Rohdewald, P., Reconstituted collagen nanoparticles, a novel drug delivery system, *J. Pharm. Pharmacol.*, 35, 537, 1983.
22. Couvreur, P., Kante, B., Lenaerts, V., Seailteur, V., Roland, M., and Speiser, P., Tissue distribution of antitumour drugs associated with polyalkylcyanoacrylate nanoparticles, *J. Pharm. Sci.*, 69, 199, 1980.
23. Kennedy, K. T., Preparation and Testing of Nanoparticles Containing Triamcinolone Acetonide, M. Pharm. thesis, Victorian College of Pharmacy Ltd., Melbourne, 1983.
24. Ehrsson, H., Lönroth, U., Wallin, I., Ehrnebo, M., and Nilsson, S. O., Degradation of chlorambucil in aqueous solution — influence of human albumin binding, *J. Pharm. Pharmacol.*, 33, 313, 1981.
25. Wosilait, W. D., Theoretical analysis of the binding of salicylate by HSA: the relationship between free and bound drug and therapeutic levels, *Eur. J. Clin. Pharm.*, 9, 285, 1976.
26. Sen, M., Mitra, S. P., and Chattoraj, D. K., Thermodynamics of binding of anionic detergents to BSA, *Ind. J. Biochem. Biophys.*, 17, 370, 1980.
27. Harrington, R. and Barry, B. W., Hydrophobic Interaction Chromatography of Bovine Serum Albumin Nanoparticles, British Pharmaceutical Conference, Edinburgh, 1982, 79P.
28. Oppenheim, R. C., Marty, J. J., and Stewart, N. F., The labeling of gelatin nanoparticles with ^{99m}technetium and their in vivo distribution after intravenous injection, *Aust. J. Pharm. Sci.*, 7, 113, 1978.
29. Oppenheim, R. C., Sberna, F., and Lichtenstein, M., unpublished data, 1983.

30. Oppenheim, R. C. and Stewart, N. F., The manufacture and tumour cell uptake of nanoparticles labeled with fluorescein isothiocyanate, *Drug Dev. Ind. Pharm.*, 5, 563, 1979.
31. Gipps, E. M., The Incorporation of Cytotoxics into Nanoparticles, M. Pharm. thesis, Victorian College of Pharmacy Ltd., Melbourne, 1983.
32. Boag, C. C., Oppenheim, R. C., Montague, P., and Birchall, R., The manufacture and in vivo testing of nanoparticles containing flukicides, *Asian J. Pharm. Sci.*, 2, 27, 1980.
33. New, R. R. C., Chance, M. L., Thomas, S. C., and Peters, W., Antileishmanial activity of antimonials entrapped in liposomes, *Nature (London)*, 272, 55, 1978.
34. Armstrong, R. D., English, J., Gibson, T., Chakroborty, J., and Marks, V., Serum methylprednisolone levels following intra-articular injection of methylprednisolone acetate, *Ann. Rheum. Dis.*, 40, 571, 1981.
35. Hashida, M., Muranishi, S., and Sezaki, H., Evaluation of water in oil and microsphere in oil emulsions as a specific delivery system of 5-fluorouracil into lymphatics, *Chem. Pharm. Bull.*, 25, 2410, 1977.
36. Morimoto, Y., Akimoto, M., Sugibayashi, K., Nadai, T., and Kato, Y., Drug carrier properties of albumin microspheres in chemotherapy. IV. Antitumour effect of single shot or multiple shot administration of microsphere entrapped 5-fluorouracil on Ehrlich ascites or solid tumour in mice, *Chem. Pharm. Bull.*, 28, 3087, 1980.
37. Widder, K. J., Senyei, A. E., and Ranney, D. F., Magnetically responsive microspheres and other carriers for the biophysical targeting of antitumour agents, *Adv. Pharmacol. Chemother.*, 16, 213, 1979.
38. Couvreur, P., Roland, M., and Speiser, P., U.S. Patent 4,329,332, 1982.
39. Gipps, E. M., Oppenheim, R. C., Forbes, J., and Whitehead, R., submitted, 1984.
40. Illum, L. and Jones, P. D. E., and Davis, S. S., Drug targeting using monoclonal antibody-coated nanoparticles, in *Microspheres and Drug Therapy: Pharmaceutical, Immunological, and Medical Aspects*, Davis, S. S., Illum, L., McVie, J. G., and Tomlinson, E., Eds., Elsevier, Amsterdam, 1984, section 4, chap. 1.

Chapter 2

BIODEGRADABLE POLYMERIC NANOPARTICLES AS DRUG CARRIER FOR ANTITUMOR AGENTS

P. Couvreur, L. Grislain, V. Lenaerts, F. Brasseur, P. Guiot, and A. Biernacki

TABLE OF CONTENTS

I. INTRODUCTION*

Entrapment of cytotoxic drugs inside endocytizable carriers such as liposomes improves the specificity of the drug and reduces its toxicity towards nondiseased cells.[1-2] Work in this field has resulted in the development of polyacrylamide nanocapsules.[3] Polyacrylamide nanocapsules may also be helpful in promoting cellular uptake via endocytosis for compounds that normally do not reach to lysosomes.[4] Owing to their polymeric nature, these small capsules (diameter of 200 nm) may be more stable than liposomes in biological fluid and during storage. Furthermore, they can entrap various molecules in a stable and reproducible way.

However, this lysosomotropic particulate carrier is unlikely to be digested by lysosomal enzymes, and this may restrict its clinical use. This is why we recently developed biodegradable nanoparticles prepared by polymerization of various alkylcyanoacrylate monomers.[5] Similar polymers are used in surgery as sutures and adhesive agents.[6]

II. PREPARATION AND CHARACTERIZATION OF POLYALKYLCYANOACRYLATE NANOPARTICLES

A. Preparation

The major advantage of alkylcyanoacrylates over other acrylic derivates which were previously used in the preparation of submicroscopic nanoparticles lies in the polymerization process of monomers. Unlike other acrylic derivates, the polymerization of which requires an energy liable to hinder the good adsorption of the drug, alkylcyanoacrylates may be polymerized without that energy. The preparation process of nanoparticles consists of adding the monomer to an aqueous surfactant solution and stirring in order to obtain micella. The pH value of this polymerization medium ranges from 2 to 4. The alkylcyanoacrylate is polymerized at room temperature and the biologically active substance is introduced into the medium either before the addition of the monomer or after the polymerization. The pH of the medium influences both the polymerization rate and the adsorption degree of the drug when this latter is in its ionized form. It is to be noted that the polymerization process is slowed down when

* Abbreviations used in this chapter are PMN, polymethylcyanoacrylate nanoparticles; PEN, polyethylcyanoacrylate nanoparticles; PBN, polybutylcyanoacrylate nanoparticles; PIN, polyisobutylcyanoacrylate nanoparticles; PHN, polyhexylcyanoacrylate nanoparticles; DACT, dactinomycin; DOX, doxorubicin; VLB, vinblastine.

the pH is reduced. Besides, the adsorption of the drug reaches a maximum when it is in a nonionized form or when the pH corresponds to the pKa value of the drug.[5]

More recently, it has been discovered, surprisingly, that nanoparticles can be prepared in an aqueous medium without requiring the presence of a surfactive agent. In some cases the presence of an acid in the polymerization medium can be avoided and is sometimes unwanted because the adsorbed drug can be adversely affected or even decomposed by the acid. The surfactive agent-free medium used for the preparation of nanoparticles generally contains polyethyleneglycol 200 (10%), glucose (5%), lactose (9.5%), or dextran 70 (1%) in an aqueous solution of citric acid. It is important to note that alkylcyanoacrylates are practically insoluble in water, and it was therefore totally unexpected that a monomer of this type, when added to a surfactive agent-free aqueous medium, could polymerize as to yield submicroscopic particles showing a diameter sometimes smaller than 100 nm. Furthermore, the fact that nanoparticles can be prepared by adding an alkylcyanoacrylate to a nonacidic aqueous medium was surprising because these monomers are very reactive and quickly polymerize by an anionic polymerization mechanism (Figure 1) as soon as they come into contact with alkaline substances or with nonacidified water.[7]

B. Molecular Weight

Polyalkylcyanoacrylate nanoparticles (PAN) were prepared with and without using surfactive agents at various pH ranging between pH 2.9 and 4.1. The molecular weights of these preparations were determined by an HPLC-GPC chromatography on a Styragel column. The molecular weight evaluations were obtained from a calibration curve of polyethyleneglycol as a standard. Figure 2 shows the chromatogram of PAN prepared at pH 3.55 in the absence of any surfactant. The presence of only one peak demonstrates that the analyzed polymer was homogeneous. Furthermore, the symmetry of the peak suggests a uniform distribution of the molecules with a calculated molecular weight ranging between 500 and 1000. Thus, nanoparticles seem to be built by an entanglement of numerous little oligomeric subunits rather than by the rolling-up of one or few long polymers. Figure 3 shows that the molecular weight of these oligomers is proportional to the pH of the preparation medium.

However, when nanoparticles were prepared by polymerization of the cyanoacrylate monomer in an aqueous medium containing Pluronic F 68 (1%) as a surfactant, the chromatogram appeared quite different (Figure 4). Indeed, three different peaks were observed: the central one corresponds to the Pluronic whereas the extreme ones correspond to a bimodal distribution of the polymeric molecular weight. The mean molecular weight of the heaviest fraction was calculated to be higher than 35,000 while the lowest one was of about 2000. These data suggest that the nanoparticle preparation process can deeply modify the molecular weight of the nanoparticle-forming polymeric subunits. These observations can be significant in the later discussion of toxicological and pharmacological results.

C. Drug Adsorption and Drug Release

Owing to their small size and porous nature, nanoparticles are able to adsorb a wide variety of drugs which are listed in Table 1. The drugs to be adsorbed onto the carrier can be dissolved in the polymerization medium before the addition of the alkylcyanoacrylic monomer or in the already prepared nanoparticle suspension.

It was important to ascertain the hypothesis of drug release from the nanoparticles as a consequence of polymer degradation. For this reason, we compared the kinetics of drug release from the carrier and with that of polymer degradation. ^{14}C nanoparticles loaded with ^{3}H dactinomycin-D (DACT) (2 mℓ) were incubated at 37°C in 50-mℓ portions of an isotonic medium buffered at various pH (acetate buffer 0.05 *M*, pH =

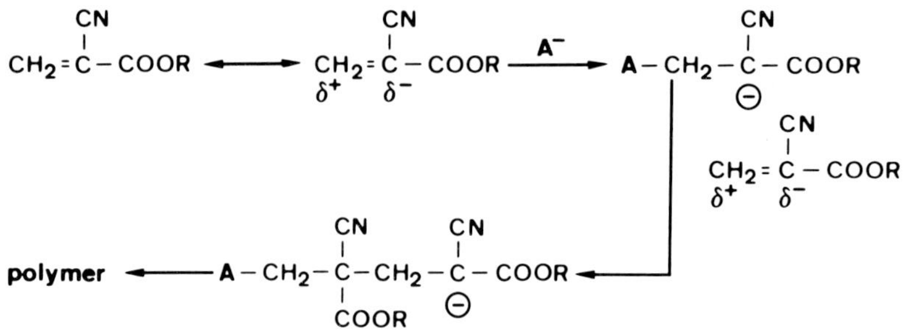

FIGURE 1. Polymerization mechanism of cyanoacrylic monomers. (From Donelly, E. et al., *Poly. Lett.*, 15, 399, 1977. With permission.)

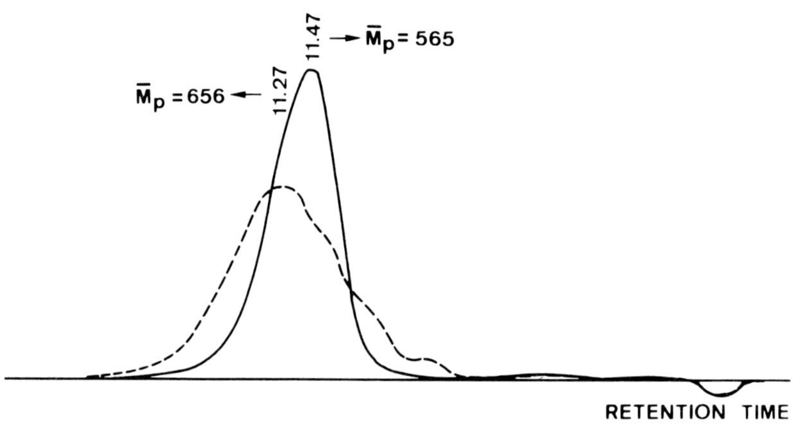

FIGURE 2. Chromatogram of PIN prepared at pH 3.55 (—) and neutralized at pH 7 (- -). (From Leyh, D. et al., *Labo Pharma*, 32, 100, 1984. With permission.)

4.9); phosphate buffer 0.05 *M*, pH = 7, and borate buffer 0.05 *M*, pH = 9). Samples (5 mℓ) were taken out at various times and centrifuged (21,000 rpm; 2 hr). ^{14}C and tritium activities were measured by liquid scintillation counting in the supernatant and the dimethylformamide-dissolved sediment. Figure 5 shows an excellent correlation between the release of ^{3}H-DACT from the nanoparticles and the degradation of the ^{14}C-polymer into water-soluble compounds. An increase in the degradation rate at higher pH values was also confirmed. The correlation between drug release and degradation rate was observed at all pH values tested. This strict correlation is only possible if the drug is distributed homogeneously in the polymeric mass of the particles. Since the biodegradability of polyalkylcyanoacrylates depends on the nature of the alkyl chain it is possible to select a product, which has a biodegradability corresponding to the program established for the drug release. The degradation kinetics of these nanoparticles as well as the release of the adsorbed drug can be perfectly controlled and programmed according to the desired therapeutic effect. This purpose can be reached by using the appropriate mixture of nanoparticles.[8]

D. Morphometrical Properties

The first observation of PAN was performed by scanning electron microscopy[5] by examination of sample shadowed in a cathodic evaporator. A gold layer about 25 nm thick was applied on the surface of the dried preparations. An homogeneous population of spherical particles with a mean diameter of about 200 nm was observed (Figure 6).

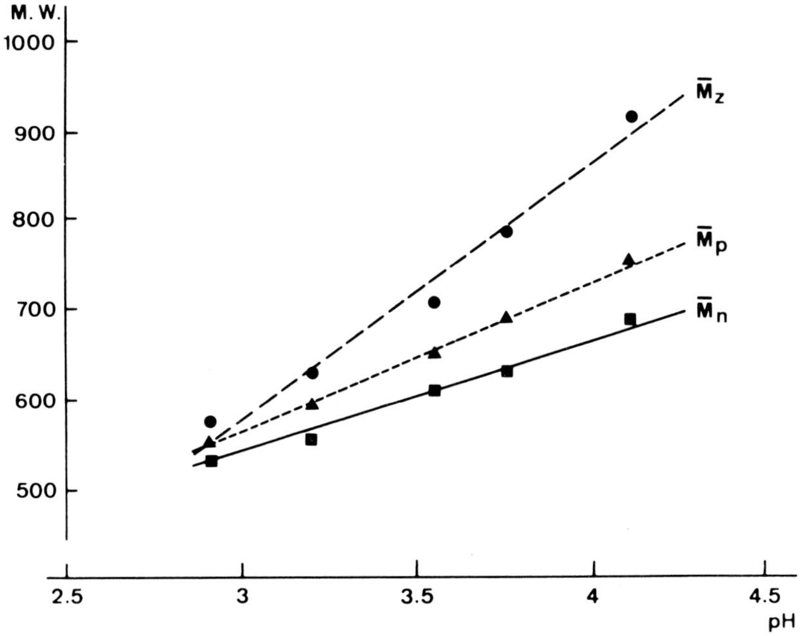

FIGURE 3. Evolution of the molecular weights as a function of the pH of nanoparticle preparation. (From Leyh, D. et al., *Labo Pharma,* 32, 100, 1984. With permission.)

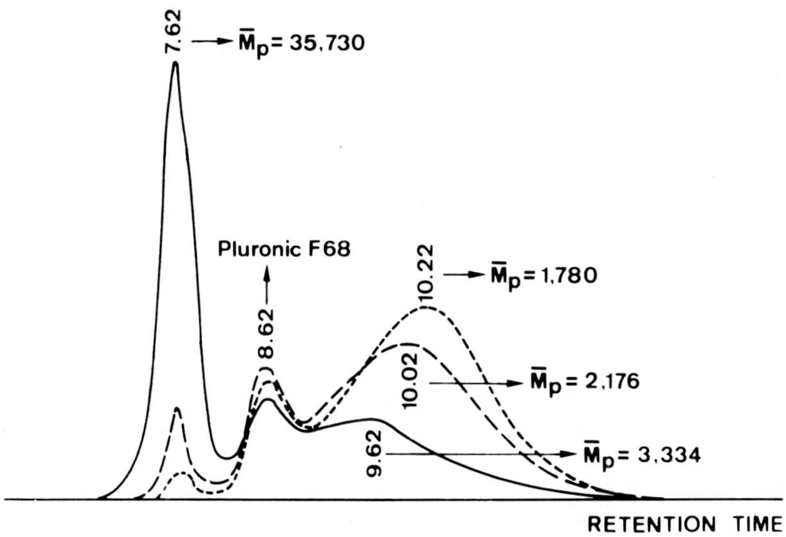

FIGURE 4. Chromatogram of PIN prepared with Pluronic F68 as a surfactive agent. pH of nanoparticles preparation: pH 4.5 (—), pH 3.8 (- -), pH 3.3 (---). (From Leyh, D. et al., *Labo Pharma,* 32, 100, 1984. With permission.)

In order to establish the morphometrical properties of the particle suspensions, these preparations were spray-frozen in liquid propane.The spraying gun was set at a 10-cm distance perpendicularly to the propane surface. The nozzle aperture was 150 μm and the nitrogen pressure was 50 kPa. Owing to the quick freezing of the sample, inclusion

Table 1
DRUG SORPTION ON POLYALKYLCYANOACRYLATE NANOPARTICLES

Drug	Concentration (μg/mℓ)	Adsorption (%) on polyalkylcyanoacrylate nanoparticles: Methyl	Ethyl	Isobutyl
Adriamycin	1000	—	—	94
Dactinomycin	55	92	86	70
Vinblastine	1000	58	66	46
Vincristine	300	—	—	57
Penicillin V	1000	50	37	—
Insulin	12 IU	40	44	40
Triamcinolone acetonide	25	—	—	44
Levamisole	800	23	28	—

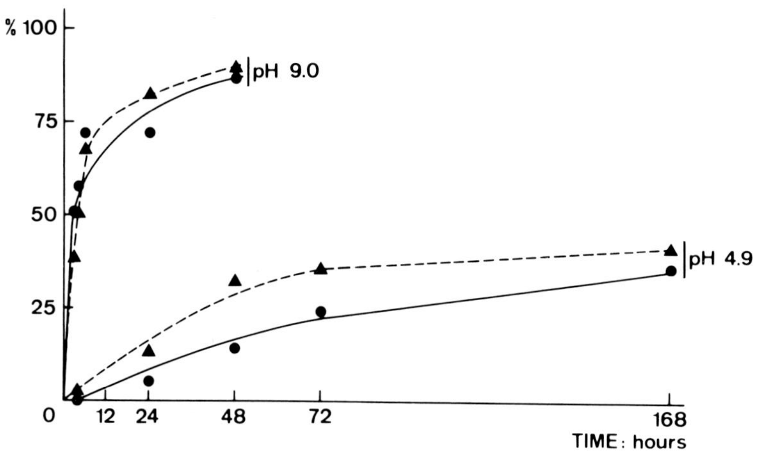

FIGURE 5. Quantities (%) of ^{3}H (▲) and ^{14}C (•) released from ^{3}H-DACT loaded ^{14}C-PIN as a function of time, in different pH conditions. (From Lenaerts, V. et al., *Biomaterials,* 5, 65, 1984. With permission.)

of cryoprotectants could be avoided. This methodology provides representative samples since it is based on the formation of an aerosol which is a random phenomenon. The frozen droplets (10 to 30 μm in diameter) recovered after propane evaporation,were embedded in butylbenzene at −88°C and freeze-fractured at −110°C according to the method of Moor et al.[9] After having shadowed the fractured surfaces with platinum (4 nm) at 45°, an 18-nm thick carbon layer was deposited at 90°. These replicas were washed in 40% chromic acid, rinsed with distilled water, placed on copper grids, and processed for electron microscopy observation. The magnifications of the electron micrographs were calibrated with a grating replica. Each square of the grid covered by the replica corresponded to several droplets of the suspension surrounded by the butylbenzene. Overlapping of micrographs was avoided and each micrograph obtained may be considered as representative of the population. An image analyzer connected on line with a computer was used for the size analysis. The size of each profile was defined as the radius of the circle with equivalent area. It must be noticed that in most cases (98%) the general shape of the profiles can be approximated to a

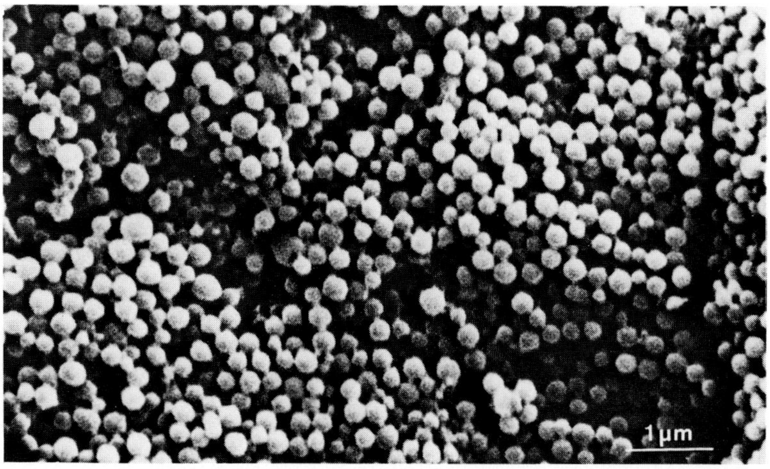

FIGURE 6. Scanning electron microscopy of PIN.

circle. If the few elongated profiles are considered as ellipses, the maximal ratio between the apparent major and minor axes is smaller than 1.5. Particle size distribution was then computed from the experimental profile size distribution using the Wicksell transformation.[10]

This method takes into account the probability for the particles to be fractured at different levels. In the profile size distribution, the class with the largest radius is entirely made of sections passing through the center of the largest particles. Since these particles form an homogeneous subpopulation of known radius, their abundance in the original population can be estimated from the abundance of their profiles in sections. Furthermore, the contribution they will give to the classes of smaller radius in the experimental profile distribution can be computed and subtracted. Thus, a new profile size distribution is obtained with one class less, on which the same procedure is applied. This leads to the construction of the particle size distribution histogram class by class from the largest to the smallest one (until all profiles have been taken into account). The application of this method requires spherical particles which is the case here. This methodology initially set up by Guiot et al.[11-13] on liposomes was demonstrated to be applicable as such for nanoparticulate vector suspensions analysis.[14-15] As an example, Figure 7 shows the general appearance of fractured polybutylcyanoacrylate nanoparticles (PBN). On the electron micrographs, particles often appear as formed of a porous core available for a high drug adsorption, surrounded by a more homogeneous ring. This latter cannot be considered as an envelope since there is no clear discontinuity between this ring and the kernal.

A particle size distribution (Figure 8a) was performed on the examination of 250 profiles and Table 2 summarizes the main morphometrical properties of the nanoparticles. In some cases, laser light scattering technique was used to determine the average size of particles. A mean radius of 276 nm was found. This value seems in contradiction with the results arising from the stereological analysis described above. However, assuming Rayleigh scattering, the measurement based on the diffusion of a laser beam by the particles is a function of the square of the particles volume and should be very sensitive to a small proportion of large particles in the population. This discrepancy is apparent in Figure 8b which represents the volume distribution of the analyzed suspension. It can so be established (from the stereological analysis) that the particles corresponding to the average volume or to the average square volume of the population have a radius of 262 or 291 nm respectively. The agreement with the value obtained

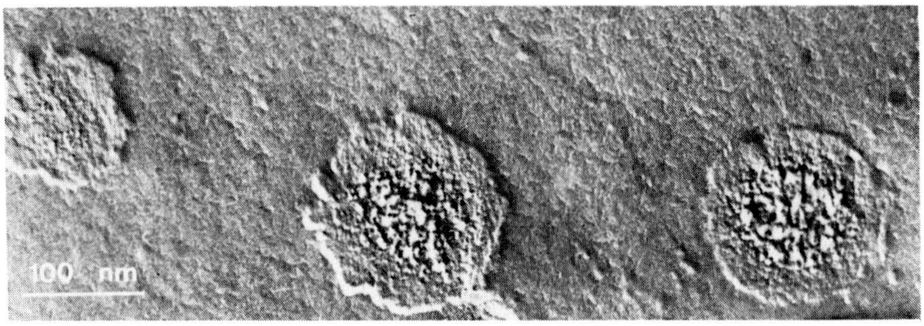

FIGURE 7. General appearance of freeze-fracture replica from spray-frozen sample of PBN suspension. Contours of profiles are often irregular. Note also the porous core surrounded by a more homogeneous ring visible in several profiles. (From Kante, B. et al., *Int. J. Pharm.*, 7, 45, 1980. With permission.)

Table 2
MORPHOMETRICAL CHARACTERISTICS OF A POLYBUTYLCYANOACRYLATE NANOPARTICLE SUSPENSION

Parameters	
Mean radius (nm)	80.6
Standard deviation for radius (nm)	73.2
Mean external surface (nm^2)	148,000
Mean total volume (nm^3)	11×10^6
Number of particles per mℓ of the suspension	1.2×10^{12}
Number of Dactinomycin molecules per particle	120

from the light scattering technique is quite satisfactory. Nevertheless, the value deduced from the stereological analysis is the one corresponding to the statistical definition of the mean. These calculations stress the influence of the measurement techniques for evaluating the parameters.

E. Degradation

1. Physicochemical Degradation

Polyalkylcyanoacrylates were extensively studied as a surgical material, with particular emphasis on their biodegradation.[16] Authors have shown the possibility for polyalkylcyanoacrylate to undergo a reverse Knoevenagel reaction producing formaldehyde and cyanoacetate (Figure 9). The kinetics of this reaction were found to be dependent solely upon physicochemical factors of the degradation medium (pH) and of the polymer itself (specific area, molecular weight and alkyl chain length). Vezin and Florence[17] have even postulated that the in vivo degradation rate was independent of any biological factor.

The production of formaldehyde by a 24-hr incubation of PIN was assessed in nonbiological media (pH 7.0 or 12.0). The reaction yielded 5% of the theoretical amount of formaldehyde (calculated for a complete reaction) in neutral medium, and 7% upon incubation in the alkaline medium. The value obtained at pH 7.0 is far too low to explain the rapid body elimination of intravenously administered nanoparticles in mice (56% elimination in one day[18]). This high clearance may be responsible for it only if nanoparticles are also degraded by another mechanism by the observation of Wade and Leonard[19] according to which, only 2 to 5% of the elimination products of an implanted polymer could be recovered under the form of formaldehyde or its metabolites.

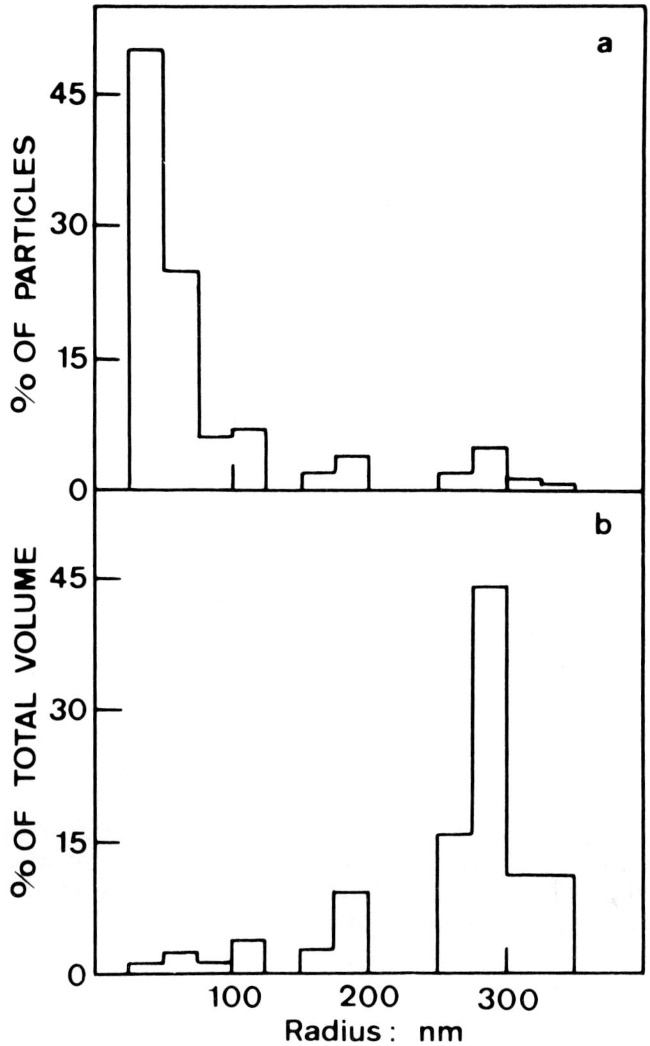

FIGURE 8. Size distribution of PBN. The upper graph gives the size distribution, on a number basis, as it is obtained from the profile size distribution in freeze-fracture replica. The histogram on the lower graph was obtained by calculating the total particle volume corresponding to each class of the upper histogram. The volume distribution was then normalized. (From Kante, B. et al., *Int. J. Pharm.*, 7, 45, 1980. With permission.)

Moreover, the slight difference in reaction at pH 7.0 and 12.0 cannot account for the tremendous difference we observed concerning the dissolution rate of nanoparticles. Indeed, most of the polymer was still present in suspension after a 24-hr incubation in the neutral medium, while only a few seconds were necessary for all nanoparticles to dissolve in the pH 12.0 medium. For these reasons, another degradation mechanism was involved in the ester hydrolysis which was likely to yield a primary alcohol and a water soluble cyanoacrylic acid polymer. The formation of isobutanol was evaluated in the above-mentioned nanoparticle preparations. The quantity of alcohol produced at pH 12.0 corresponded to 85% of the theoretical amount which would have been released by a complete reaction. This reaction can therefore be con-

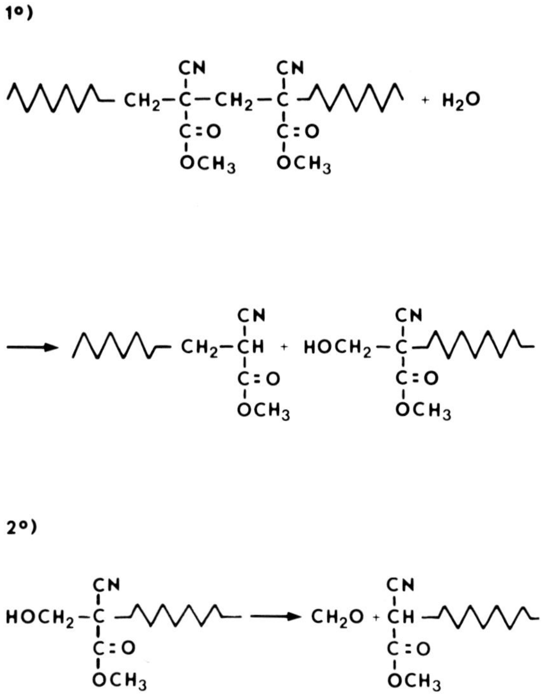

FIGURE 9. PIN degradation mechanism.

sidered as responsible for the observed dissolution of the nanoparticles in the alkaline medium.

In another experiment, where the quantities of NaOH used to hydrolyze the ester functions of PIN at pH 10.5 were recorded for 4 hr, the total amount of delivered NaOH corresponded to the hydrolysis of 94% of the initially present ester functions. In this experiment we could estimate that all nanoparticles were dissolved when about half of the ester functions had been hydrolyzed (Figure 10).

2. *Biological Degradation*

Unlike the Knoevenagel reverse reaction, the degradation of PAN by ester hydrolysis was likely to be catalyzed enzymatically. This possibility was assessed in a number of in vitro experiments involving the use of several hepatic subcellular fractions in the rat.

a. *Tritosomes*

Nanoparticles were initially developed as a potential lysosomotropic carrier.[4] This led us to first choose a lysosomal fraction. ^{14}C-PHN were incubated at pH 5.6 in the presence of tritosomes (prepared according to Trouet[20]). Samples taken at various time intervals were centrifuged and the radioactivity was evaluated in the supernatant (soluble polymer) and the pellet (insoluble polymer), allowing the calculation of the percentage of polymer solubilization. Figure 11 shows that in an acid medium (pH 5.6), tritosomes have raised the degradation rate of the nanoparticles. The reaction tends to slow down after about 1 hr. Upon addition of fresh tritosomes (after 2 hr), the kinetics increase markedly indicating that the previous observed plateau might be due to a relative instability of the enzyme involved in nanoparticle solubilization.

b. *Microsomes*

The lysosomotropic nature of a carrier does not exclude the possibility of a contact with other subcellular fractions. That is why we studied nanoparticle degradation in

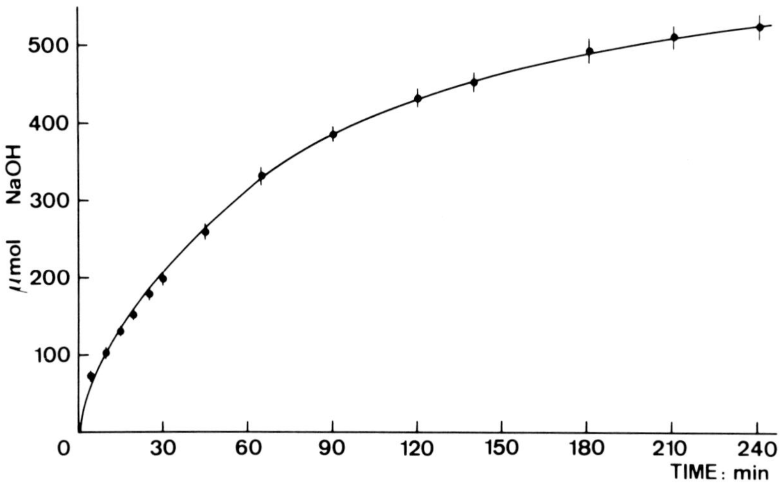

FIGURE 10. Quantities of NaOH necessary to thwart the acidity resulting from polymer degradation and to keep the pH value a constant (10.5) as a function of time. Each point is the mean of 3 experiments ± SD. (From Lenaerts, V. et al., *Biomaterials,* 5, 65, 1984. With permission.)

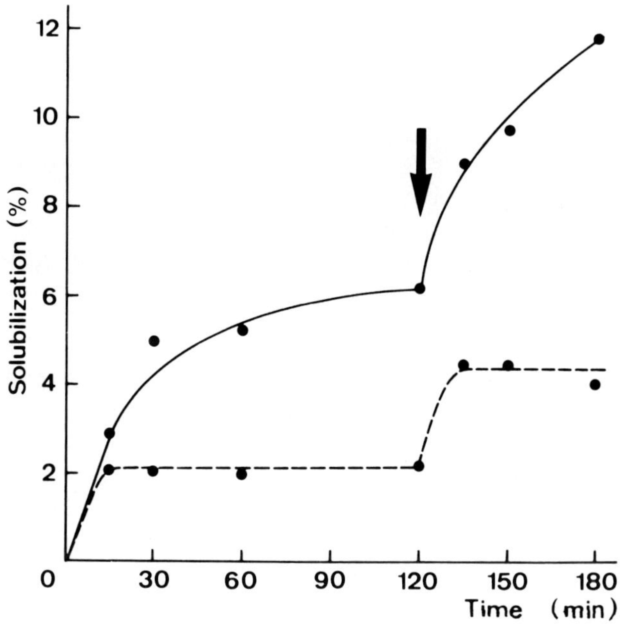

FIGURE 11. Solubilization of PIN in the presence (—) and in absence (---) of tritosomes. The arrow indicates addition of a new tritosomes aliquot.

the presence of microsomes, which are known as the major metabolization sites of xenobiotic substances. In these experiments, however, we no longer examined the solubilization of the polymer but the production of primary alcohol by the ester hydrolysis. Two experiments were undertaken with PIN. In the first one, the protein concentration was a constant (1.88 mg/mℓ), and isobutanol production was plotted against time. In the second experience, different protein concentrations were used with a constant incubation time (30 min). Figures 12 and 13 show a direct correlation between alcohol production and time (at a constant protein concentration) or protein concentration (with a constant incubation time). However, in both cases, the reaction reaches a plateau when about 5% of the ester functions were hydrolyzed. The reasons thereof have been discussed earlier.

c. Liver Extracts

Finally, the spectrum of our study was broadened by the use of a liver extract containing all subcellular fractions except the nucleus. In these experiments, the solubilization of the polymer was followed as a function of time in the presence or absence of the extract at different pH values. Two different polymers were used, namely polyisobutyl and polyhexylcyanoacrylate. Figures 14 and 15 show that at all tested pH values, the solubilization of PIN and PHN was accelerated by the presence of the extract.

In another experiment, we found no difference in solubilization rate in the absence or the presence of a heat-denatured liver extract which confirms that the part played by this biological fraction in nanoparticle degradation is a purely enzymatic mechanism. Although it yields the same products, the enzymatic degradation shows radically different characteristics than the chemical degradation as shown in Figure 16 in which the difference in the optimum activity pH appears clearly. Indeed, in the presence of the extract, the optimal pH values range between 5 and 6, while in a nonbiological medium, the reaction rate raises with the pH value.

Finally, the high degree of solubilization in the presence of the extract as compared to the tritosomes tends to prove that another subcellular fraction than lysosomes must be relatively highly active in the catalysis of nanoparticle degradation. This is even more evident when bearing in mind that only 2% of the protein contained in the liver can be attributed to lysosomes. From the results, we now have a better understanding of in vivo nanoparticle degradation. It is reasonable to assume that nanoparticles administered in vivo are degraded mainly by an enzymatic mechanism, into water-soluble polymers of such a molecular weight that they are quickly eliminated in the urine and the feces.[21,22] The low toxicity of the degradation products as compared to formaldehyde and cyanoacetate acid could explain the low toxicity that was observed in the subacute toxicity test described below.

III. INTERACTION OF POLYBUTYLCYANOACRYLATE NANOPARTICLES WITH CELLS IN CULTURE

The fate of the nanoparticles, when in interaction with cells, may be followed by adsorbing a radiolabeled tracer (^{3}H-DACT on their internal and external surfaces). In the case of PBN, the extent of such a labeling was estimated by the stereological analysis. This procedure gives the number of observed profiles per unit area of the micrographs which is correlated to the number of particles per unit volume of the suspension. In the preparation used and purified in view to remove free ^{3}H-DACT, 1.2×10^{12} particles per milliliter were present while radiolabeled concentration was 2.7×10^{-4} mol/mℓ. It becomes then very easy to compute that, on the average, one particle contains 120 DACT molecules (see Table 2). A quantitative study of the interaction between PBN and cells in culture can now be attempted on a number basis. Indeed, one can assume that the dissociation between DACT and the particles in the culture medium is not an instantaneous process, since the penetration of the medium into the internal lattice of the particles would be rather slow. In fact, the butyl moieties confer a partial hydrophobic character to the polymeric lattice. In conclusion, one can reasonably assume that the DACT uptake by cells represents also the particles uptake.

In view of the quantitative evaluation of the parameters describing cell-particle interaction, we have chosen mouse peritoneal macrophages in culture as a model. Peri-

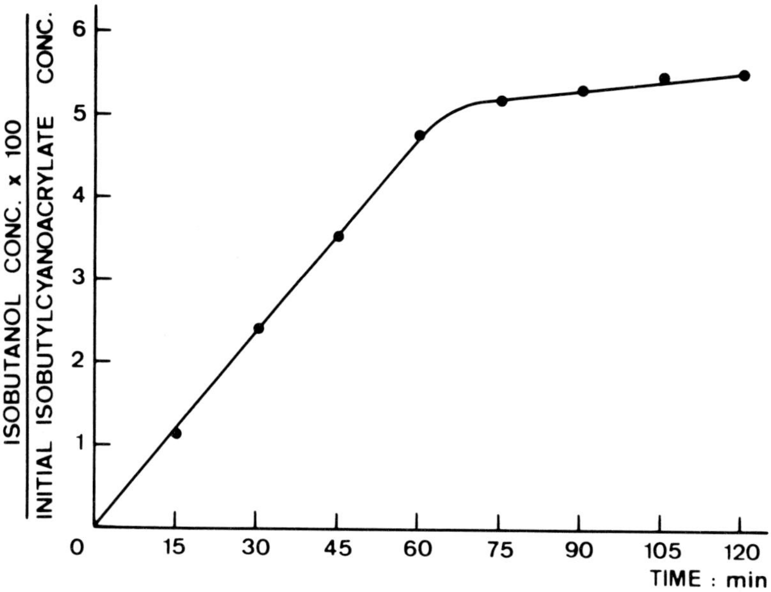

FIGURE 12. Isobutanol produced by ester hydrolysis of PIN in the presence of rat liver microsomes (1.88 mg protein/mℓ), as a function of time. (From Lenaerts, V. et al., *Biomaterials*, 5, 65, 1984. With permission.)

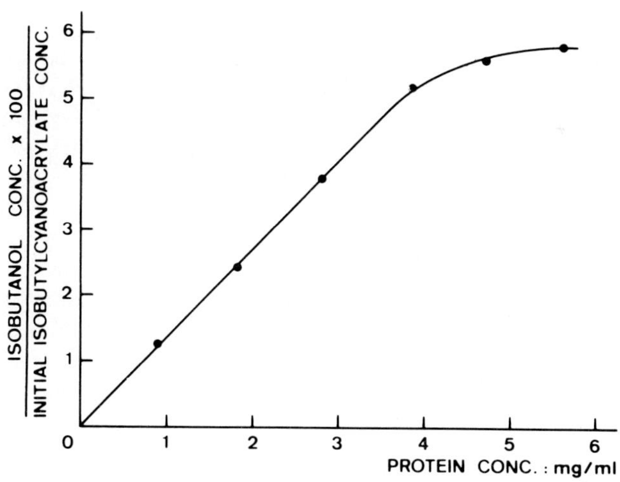

FIGURE 13. Isobutanol produced by ester hydrolysis of PIN in 30 min in the presence of different concentrations of rat liver microsomes. (From Lenaerts, V. et al., *Biomaterials*, 5, 65, 1984. With permission.)

toneal cavity of C57BL mouse was washed three times with Eagle Dulbecco medium. The collected fluid from one mouse was introduced in a Petri dish of 6 cm diameter. After 2 hr of incubation at 37°C, the cells still in suspension were removed and the adherent macrophages were washed twice with a phosphate-buffered saline solution and cultured in 5 mℓ Eagle Dulbecco medium either containing or not (depending on the experimental conditions) 20% fetal calf serum. The atmosphere of incubation was enriched in CO_2, the concentration of which was maintained at 10%. After incubation

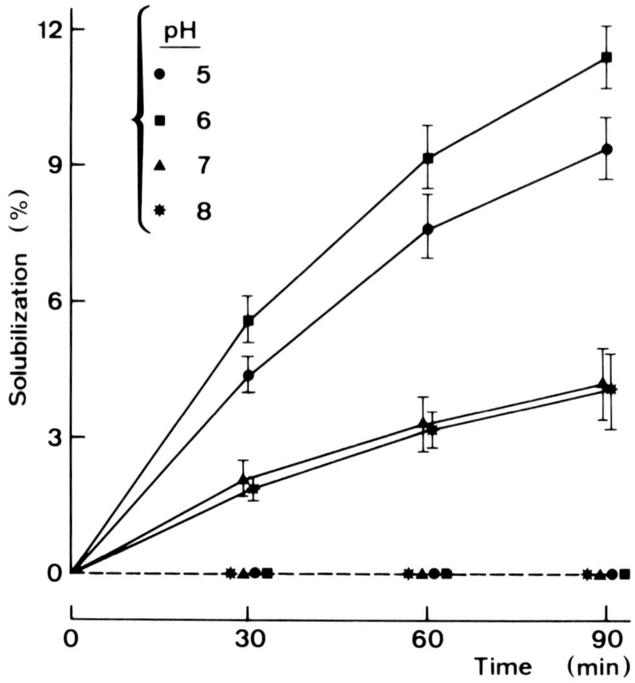

FIGURE 14. Solubilization of ^{14}C-PHN in the presence (—) or absence (---) of rat liver extracts at 37°C and for various pH.

in the presence of particles, at various time intervals or at different temperatures, the cells were washed once with the culture medium and three times with the phosphate buffered saline solution. After homogenization, cell protein concentration was determined by the method of Lowry et al.[23] and the radioactivity of ^{3}H-DACT was measured by liquid scintillation counting.

In a first experiment we had confirmed the high potentialities of the nanoparticles to release drugs into cells. We have compared the uptake of ^{3}H-DACT by mouse peritoneal macrophages when this compound was free in the culture medium or adsorbed on the nanoparticles, the extracellular concentration of ^{3}H-DACT being the same in both cases. It is shown in Figure 17 that the radiolabeled uptake is three times more important when the compound is presented to cells by way of the vectors (carriers) and when 20% fetal calf serum is present in the culture medium (comparison between blocks 1 and 3 of Figure 17).

After having demonstrated the value of the nanoparticles as an intracellular drug carrier, we have determined some parameters characteristic of the kinetics of the cell interaction. In the kinetics of association as shown in Figure 18, the number of DACT molecules bounded per milligram cell protein can be converted into the number of particles associated per cell since we know that one particle contains on the average 120 drug molecules and that one cell corresponds to 83.3 pg of cell protein.[24] Table 3 gives the main characteristics related to this kinetics of interaction. During this interaction, a plateau is reached representing roughly 120 nanoparticles per cell (Figure 18). This number corresponds to a volume equivalent to 0.34% of the cell volume itself but also to a delivery of 120 × 120 = 14,400 DACT molecules per mouse peritoneal macrophage.

In order to get some information on the mode of interaction we also followed the effect of the temperature on the nanoparticle-cell association. Figure 19 shows that this

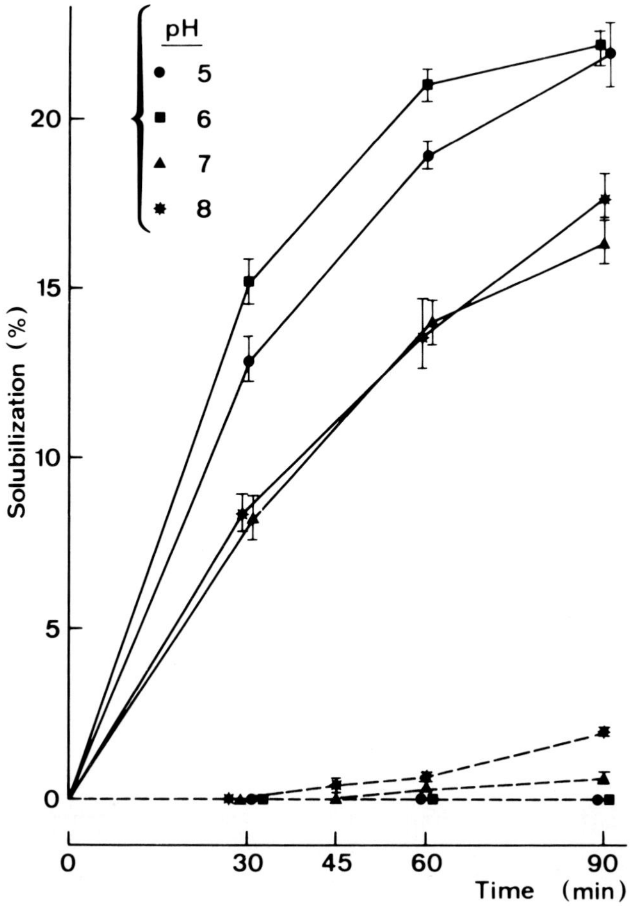

FIGURE 15. Solubilization of ^{14}C-PIN in the presence (—) or absence (---) of rat liver extracts at 37°C and for various pH.

association is nearly nonexistent around 0°C and is rapidly increasing when the temperature raises up to 20°C. This suggests that the association is cell-energy dependent. The interaction is also dependent upon the presence of serum in the incubation medium (blocks 2 and 3 in Figure 17). Since the polymeric nature of the particles excludes fusion with the plasma membrane, the pattern of Figures 18 and 19 suggests that endocytosis may play a major role in the uptake process.

Scanning electron microscopy observations of mouse peritoneal macrophages interacting with nanoparticles provide us with additional evidence for endocytosis. Indeed, the uptake of the particulate carriers seems to enhance the microvilli formation by cells. Figure 20 shows clearly that the particles are taken up by means of these microvilli before being internalized into the cells.

IV. TISSUE DISTRIBUTION AND PHARMACOKINETICS OF POLYALKYLCYANOACRYLATE NANOPARTICLES AND ASSOCIATED DRUGS

The exact knowledge of the body distribution of nanoparticles appears to be essential for the understanding of variations in toxicological and pharmacodynamic parameters of targeted drugs. From a toxicological point of view, the targeting of drugs to

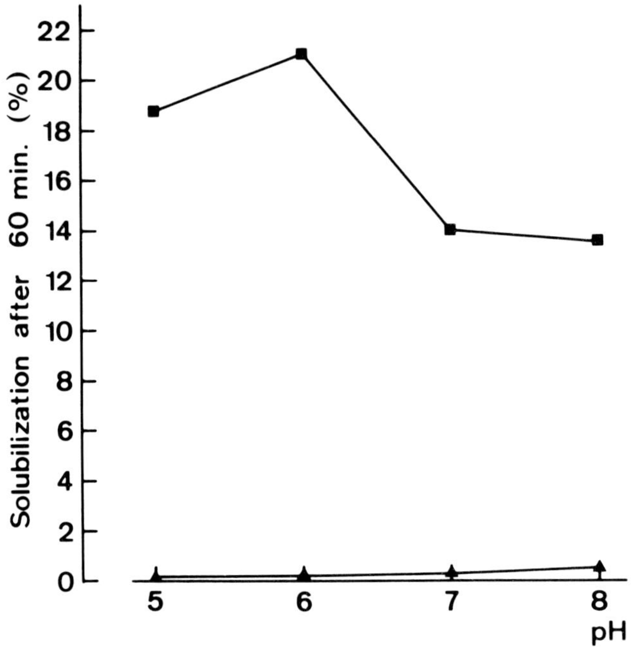

FIGURE 16. Solubilization of ^{14}C-PIN after incubation (60 min) in the presence (■) or in absence (▲) of rat liver extracts as a function of the pH.

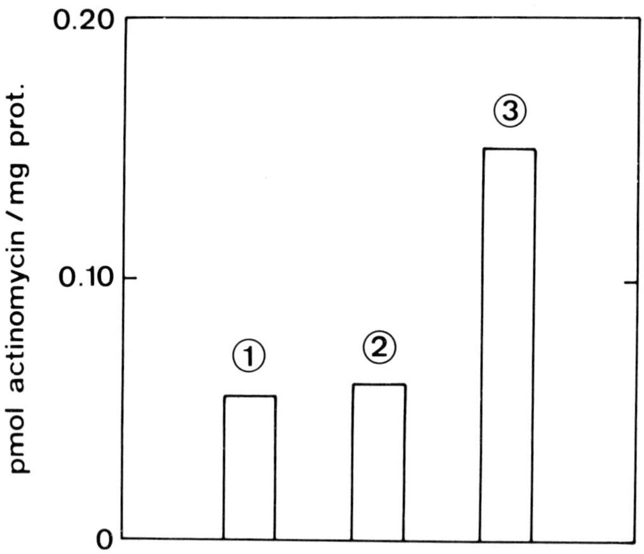

FIGURE 17. Uptake of ^{3}H-DACT by mouse peritoneal macrophages. Block 1: DACT was presented during 1 hr to the cells maintained at 37°C and cultured in a culture medium containing 20% serum. Concentration of DACT was 1.3 pmol/mℓ. Block 2: same experimental conditions as in block 1 except that the culture medium did not contain serum and DACT was adsorbed on PBN. Block 3: same experimental conditions as in block 1 except that DACT was adsorbed on PBN and presented to cell in presence of serum.

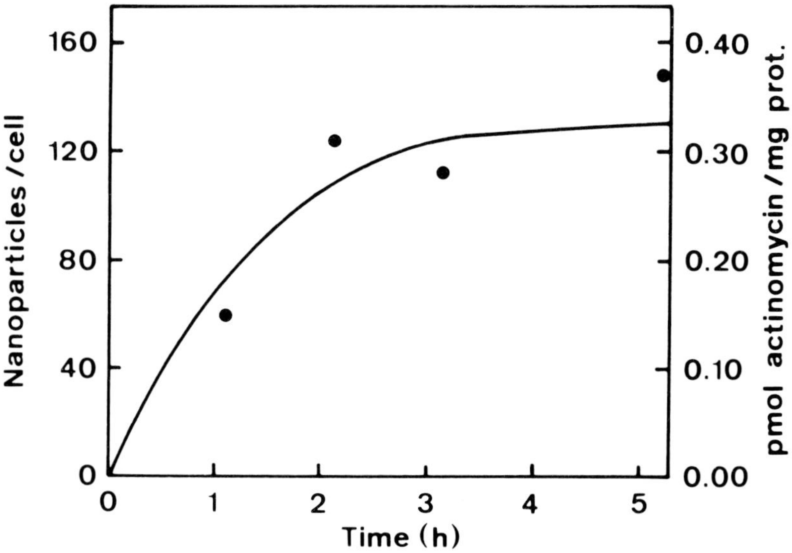

FIGURE 18. Time dependence of the association between PBN and mouse peritoneal macrophages. The external particle concentration was 1.2×10^{10}/mℓ and about 200,000 cells/mℓ were used and maintained at 37°C for different times.

Table 3
INTERACTION BETWEEN POLYBUTYLCYANOACRYLATE NANOPARTICLES AND MOUSE PERITONEAL MACROPHAGES[a]

Parameters	
Number of particles taken up by 1 cell in 1 hr	60
Corresponding vol represented by this number	0.66 μm^3
% Of the cell volume represented by this number[b]	0.17%

[a] The cells were maintained at 37°C and the external nanoparticle concentration was 1.2×10^{10} $m\ell^{-1}$.

[b] The volume represented by a mouse peritoneal macrophage being of 395 μm^3.

specific maps raises several questions. In fact, the accumulation of piloted products in tissues where the drug is toxic will cause a decrease in its therapeutic index. Conversely, the preservation of drug accumulation in tissues sensitive to its side effects will be beneficial. These few examples show how much the knowledge of the carrier's body distribution is essential to predict the influence of targeting on toxicological properties of piloted drugs. From a theoretical view point, the body distribution pattern of nanoparticles could be in close relation with their action mechanism. The results obtained by Brasseur et al.[53] showing an increase in the efficacy of adsorbed drugs against experimental tumor, may be due to several factors: the drug-nanoparticle association can impact a programmed release profile to the cytostatic drug and then can improve its tumoral bioavailability, nanoparticles could concentrate the drug specifically into tumor cells, thus enhancing the cytostatic effect.

The first part of the previous paragraph deals with the body distribution of the carrier itself. Elements such as the influence of the way of administration, the chemical

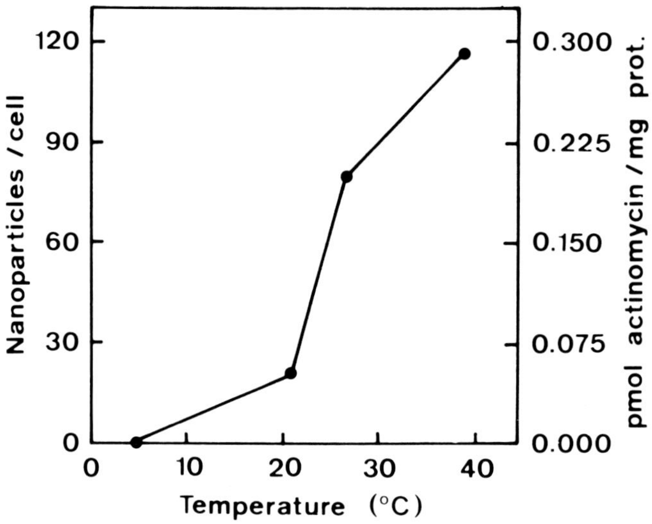

FIGURE 19. Effect of temperature on the interaction between PBN and mouse peritoneal macrophages. The external particle concentration was $1.2 \times 10^{10}/m\ell$ and about 200,000 cells/mℓ were used and incubated for 2 hr at different temperatures.

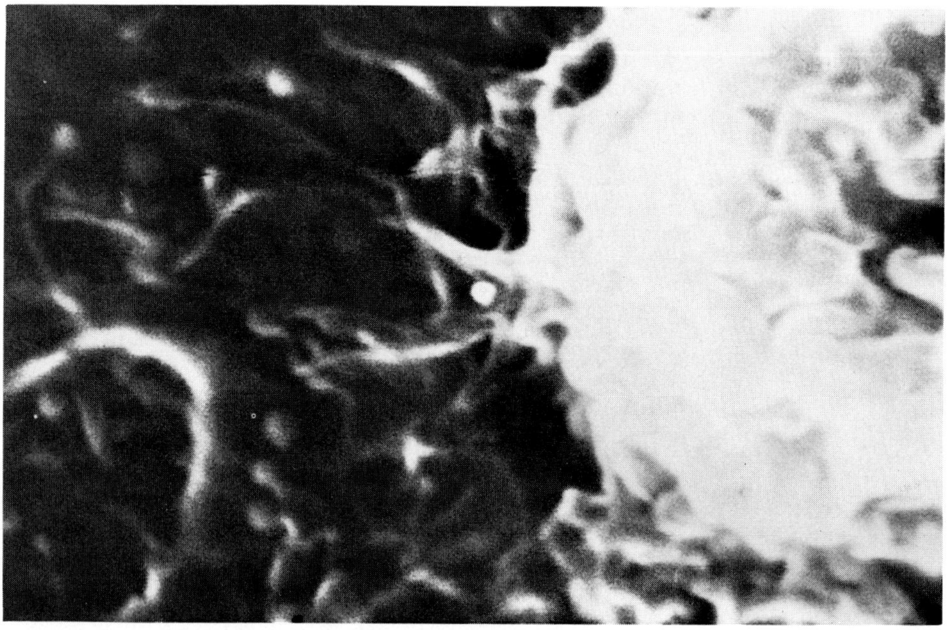

FIGURE 20. Scanning electron microscopic appearance of a macrophage interacting with nanoparticles. We see below, the cell body and above, the microvilli surrounding the particles. At the center of the graph two microvilli enveloping a particle are clearly apparent. The diameter of this particle is 200 nm. The spherical structures taken up by the microvilli are related to nanoparticles since they are not observed for macrophages in absence of particles.

nature of the polymer, and the size of the particles are discussed regarding their influence upon the distribution of the carrier in the body of healthy animals. The localization of nanoparticles in tumor-bearing animals is also examined.

The modifications in the distribution pattern of adsorbed cytostatics are reviewed in the second part of this chapter.

A. Nanoparticles in Healthy Animals

1. Intravenous Administration

a. Influence of Physicochemical Properties

The body distribution of nanoparticles has been studied using whole body autoradiography after Ullberg.[25] This technique shows radioactive material distribution without separation of organs. After administration of radioactive material, animals were frozen in liquid nitrogen. Sagittal sections 50 μm thick were taken and applied on a radiosensitive film. After exposure for 1 to 4 weeks the film was processed and the autoradiograms were obtained. The dark areas correspond to high radioactive concentrations. As an aid to the reader, position of the organs as they appear on whole-body sections is shown in Figure 21.

Nevertheless, the qualitative data given by autoradiography were further confirmed in terms of quantities by liquid scintillation counting in selected organs.

Five minutes after intravenous administration of radioactive ^{14}C-PIN, radioactivity mainly concentrated in the liver, the spleen, and the lungs (Figures 22 and 23). This phenomenon can be attributed to the colloidal nature of the carrier, which is engulfed by the reticuloendothelial cells. Thirty minutes after administration, the distribution pattern remained similar. Then intestinal and renal excretion started (Figures 22 and 23). Twenty-four hours after the intravenous administration, the body was almost free of radioactivity (Figures 22 and 24). This lack of prolonged accumulation of PIN is interesting from a toxicological view point. Side effects due to a prolonged accumulation of polymeric material in the body are likely to be avoided. The rapid clearance of PIN from the body after intravenous administration could be in close relation with the low molecular weight of the polymer as earlier discussed.

Couvreur[8] showed that the degradation rate of nanoparticles is highly dependent upon the length of the alkyl chain. It was therefore interesting to evaluate the incidence of this parameter in vivo. For this purpose, we have compared the body distribution of PIN and PHN in mice after intravenous injection. As shown on Figure 25, the distribution pattern of PHN is not different from that of PIN five minutes after administration. However 24 hr after administration (Figure 26), PIN are almost completely cleared while PHN still remain highly concentrated in the liver and the spleen.

As shown in Figures 23 and 24, the quantitative determination of body distribution of PIN and PHN confirms the data obtained by autoradiography. Five minutes after the administration, an intensive hepatic uptake was observed for both types of nanoparticles. The distribution pattern was similar for PHN and PIN during the first 4 hr. Nevertheless major differences appeared 24 hr after the administration. The hepatic uptake of PHN was twice as high as that observed for PIN. The quantitative determination of excreted radioactivity in urine and feces for both preparations (Table 4), shows that more than 90% of PIN radioactivity was excreted within a week while an excretion of only 49% was observed in the case of PHN. For both kinds of nanoparticles the urinary excretion is prevailing. This behavior could be attributed to the production of soluble degradation products able to be cleared through the renal filter.

Because of the importance of blood level on the pharmacological properties of a drug, it appeared essential to evaluate the blood clearance of the carrier after intravenous administration in mice.

Experiments were performed using PHN and PIN. As presented in Figure 27, blood clearance of PIN is faster than for that of PHN. The disappearance of radioactivity

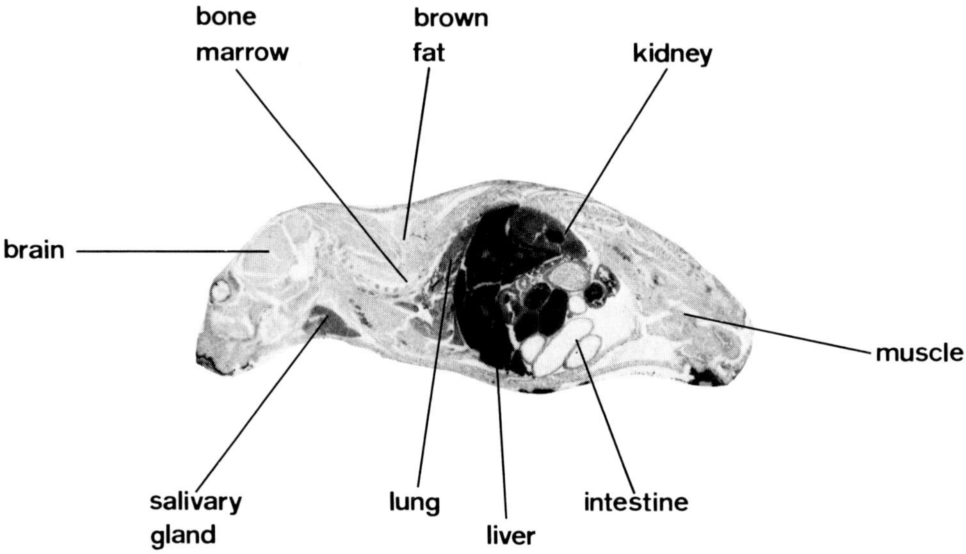

FIGURE 21. Whole body autoradiography of a mouse, 30 min after intravenous administration of ^{14}C-PIN.

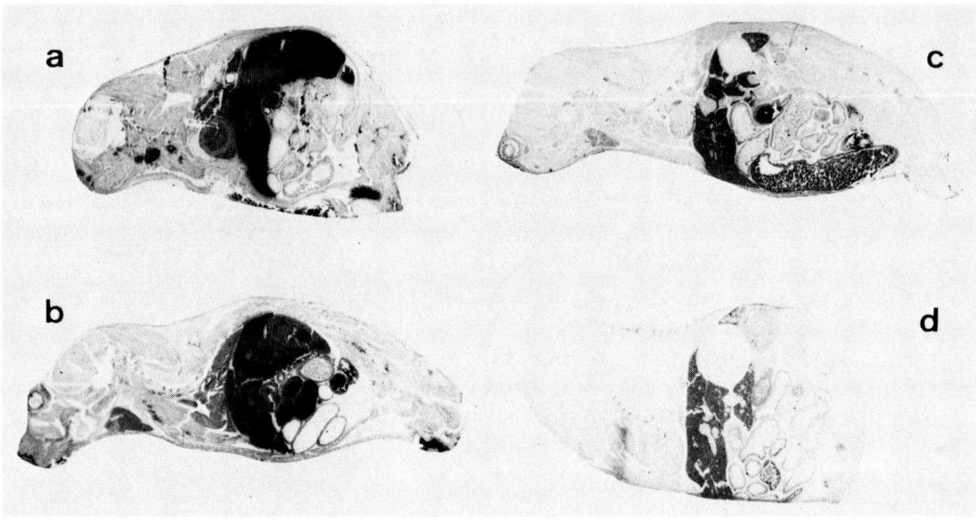

FIGURE 22. Whole body autoradiography of a mouse after intravenous administration of ^{14}C-PIN; a = 5 min, b = 30 min, c = 4 hr, d = 1 day. (From Grislain, L. et al., *Int. J. Pharm.*, 15, 335, 1983. With permission.)

from blood after injection follows 1-exponential and 2-exponential models for PIN and PHN respectively. After computing the data following a nonlinear regression model, pharmacokinetic parameters were determined for both kinds of nanoparticles (Table 5). These parameters confirm the previous observation that the heavier polymer (PHN) is cleared more slowly from the blood than its lighter homologue (PIN). These results prompt the possibility of adapting the blood clearance of the carrier by simply selecting the appropriate polymer.

The size of the carrier was also considered as a factor likely to influence the distribution in the body. Indeed, the size of the endothelial pores could limit the extravascular diffusion of the carrier and therefore the availability of the drugs towards malig-

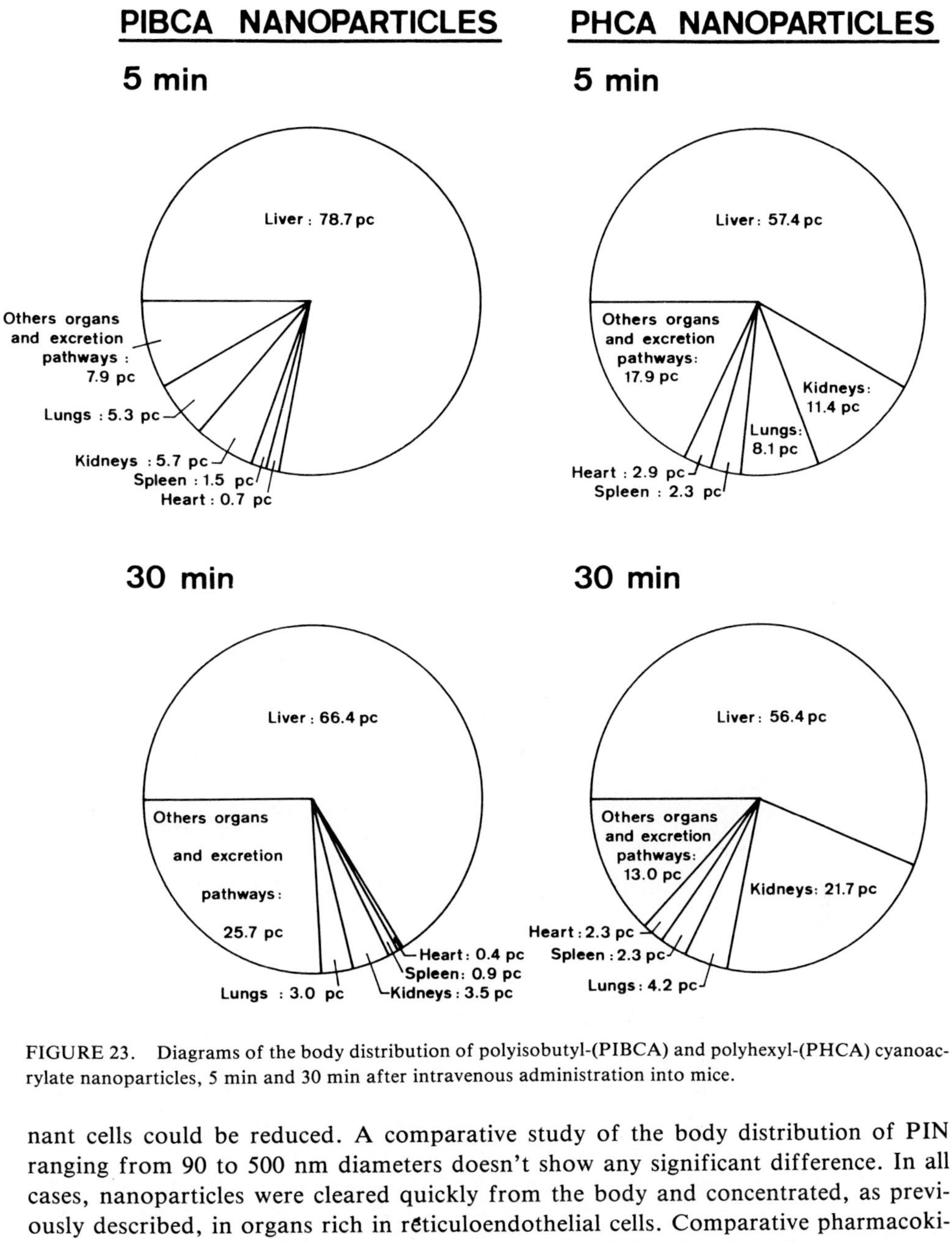

FIGURE 23. Diagrams of the body distribution of polyisobutyl-(PIBCA) and polyhexyl-(PHCA) cyanoacrylate nanoparticles, 5 min and 30 min after intravenous administration into mice.

nant cells could be reduced. A comparative study of the body distribution of PIN ranging from 90 to 500 nm diameters doesn't show any significant difference. In all cases, nanoparticles were cleared quickly from the body and concentrated, as previously described, in organs rich in reticuloendothelial cells. Comparative pharmacokinetic studies also revealed no differences between particles of various size.

Other attempts have been made in order to reduce the hepatic uptake of the carrier. We will mention reticuloendothelial blockade by high molecular weight dextran. Mice were injected with 5000 sulfate dextran. After 2 hr they were injected with radioactive PIN. Figure 28 shows that in that case, hepatic uptake can be drastically reduced. This behavior is further evidence that reticuloendothelial cells are deeply involved in the hepatic uptake of the carrier.

b. Intrahepatic Distribution

Owing to the high proportion of the carrier taken up by the liver, it is also probable that this organ represents the major metabolization site for these polymeric particles.

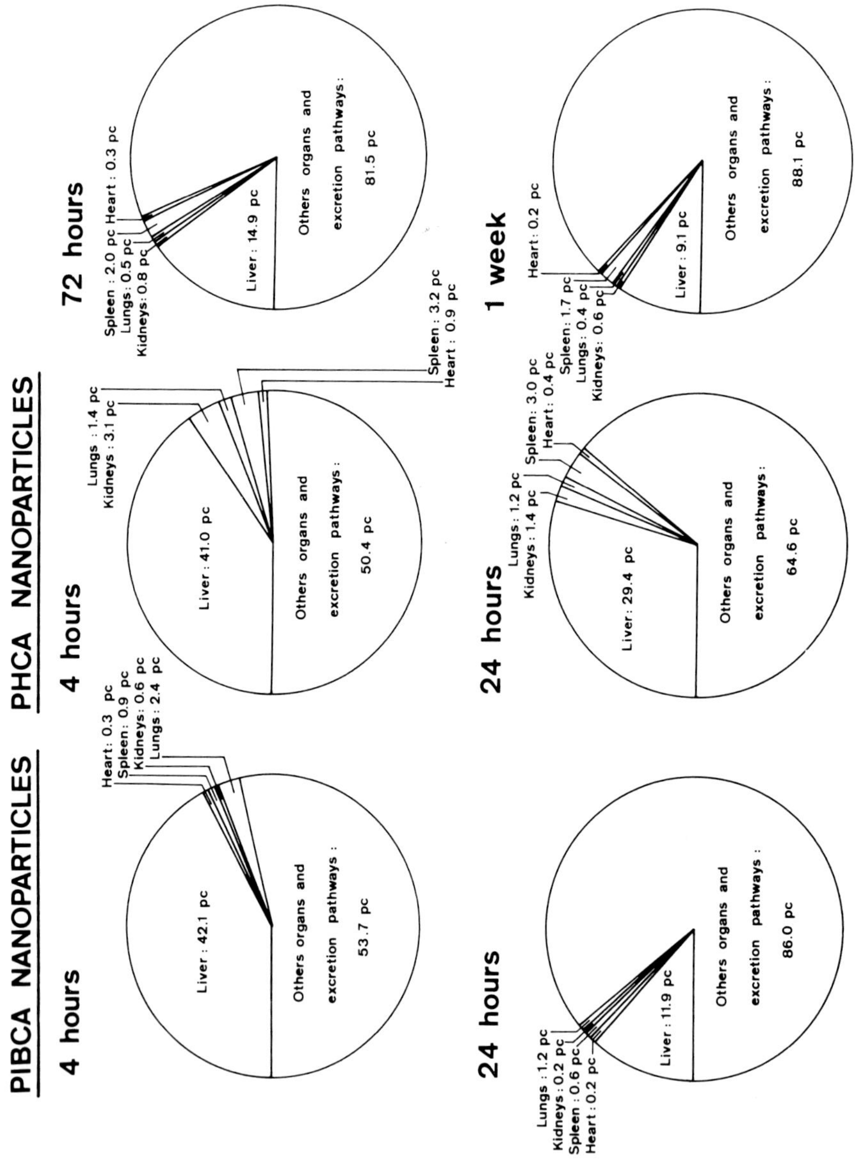

FIGURE 24. Diagrams of the body distribution of polyisobutyl-(PIBCA) and polyhexyl-(PHCA) cyanoacrylate nanoparticles, 4 hr, 24 hr, 72 hr, and 1 week after intravenous administration into mice.

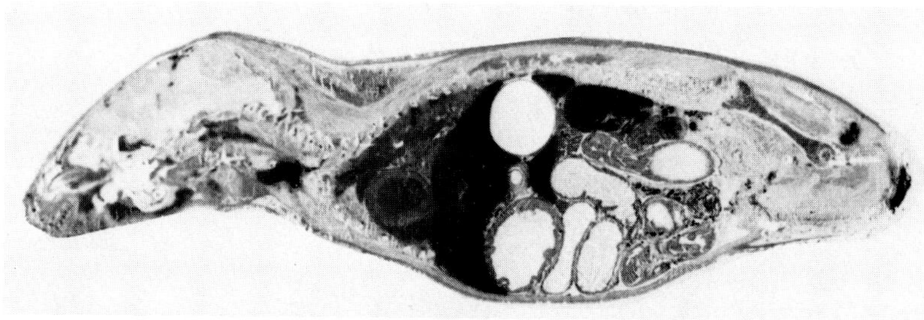

FIGURE 25. Whole body autoradiography of a mouse 5 min after intravenous administration of ^{14}C-labeled PHN.

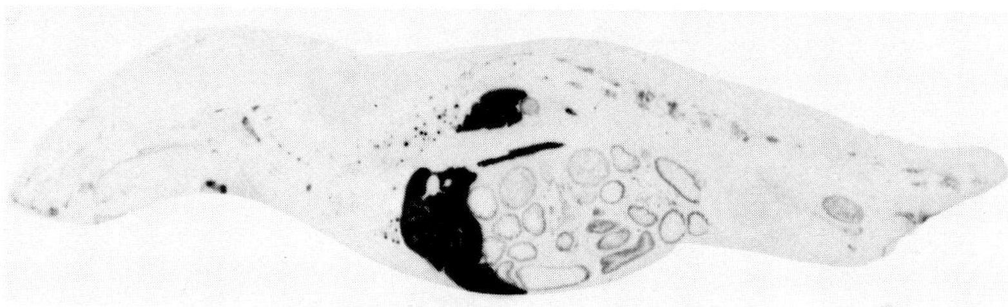

FIGURE 26. Whole body autoradiography of a mouse 24 hr after intravenous administration of ^{14}C-labeled PHN.

Table 4
URINARY AND FECAL EXCRETION (% INJECTED DOSE) OF ^{14} C-LABELED POLYISOBUTYL-(PIN) AND POLYHEXYL-(PHN) CYANOACRYLATE NANOPARTICLES

Time (days)	Urinary excretion (%)		Fecal excretion (%)	
	PIN	PHN	PIN	PHN
1	39.9	28.5	16.2	4.3
3	56.4	33.6	23.9	6.4
7	66.6	35.9	27.0	13.0

It was therefore very important to determine the distribution of nanoparticles between the different liver cell types in vivo. After intravenous administration of ^{14}C-labeled PIN to anesthetized rats, liver was perfused first with a Ca^{++} free Hank's medium and then, after removal of a liver lobe, with a pronase or collagenase solution at 8°C following a previously described method.[26] The pronase perfusion allows the collection of Kupffer and endothelial cells, purified from other cell types after elutriation.
Similarly, the collagenase solution permits the collection of parenchymal cells. Radioactivity in the cell preparations, nanoparticle samples, blood samples, and total liver fraction was determined by liquid scintillation counting after oxidation of the samples. Protein content was determined by the method of Lowry et al.[23] using BSA as a standard. Figure 29 shows that nanoparticles accumulated mainly in the Kupffer cells 30 min after injection. Indeed, a ratio of 101 was found between carrier concentrations in

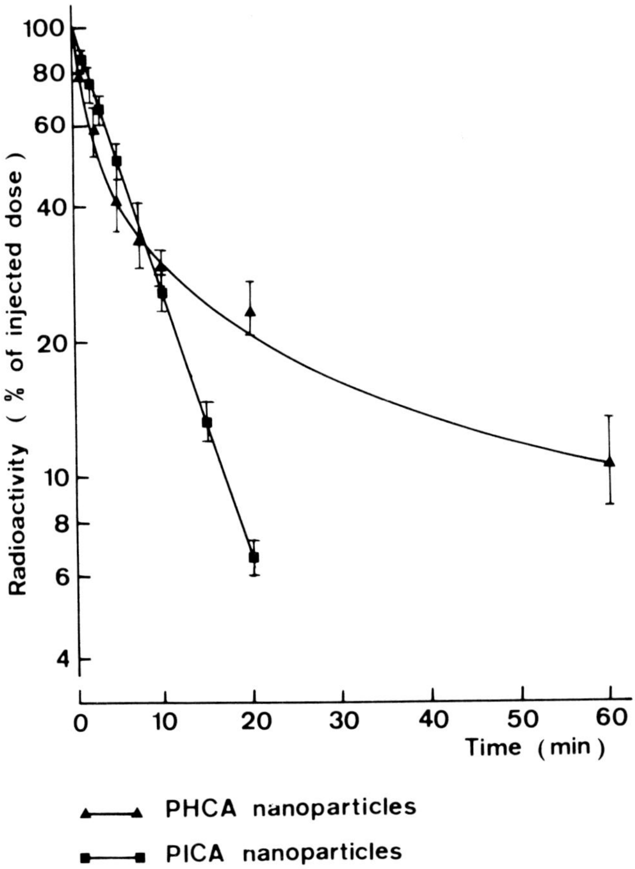

FIGURE 27. Disappearance of radioactivity from blood after injection of ^{14}C-labeled PICA and PHCA-cyanoacrylate nanoparticles into mice.

Table 5
PHARMACOKINETIC PARAMETERS OF THE BLOOD CLEARANCE OF POLYISOBUTYL-(PIN) AND POLYHEXYL-(PHN) CYANOACRYLATE NANOPARTICLES[a]

	A_o	A^1	B_o	B^1	t 1/2 α min	t 1/2 β min
PIN	9.83	−0.134	—	—	5.37	—
PHN	50.5	−0.155	36.12	−0.0011	4.47	63.0

[a] $C = A_o \cdot e - A_1 t + B_o \cdot e - B_1 t.$

Kupffer and parenchymal cells. A ratio of 4.2 was found between Kupffer and endothelial cells. This can only be explained by an intense phagocytotic capture of the carrier by the Kupffer cells.

However, the high proportion of nanoparticles observed in the liver in both autoradiography and body distribution studies was not found in this experiment.

A study of total liver radioactivity during a Ca^{++}-free perfusion has shown that half of the nanoparticles were gradually flushed out. This can be accounted for by the fact

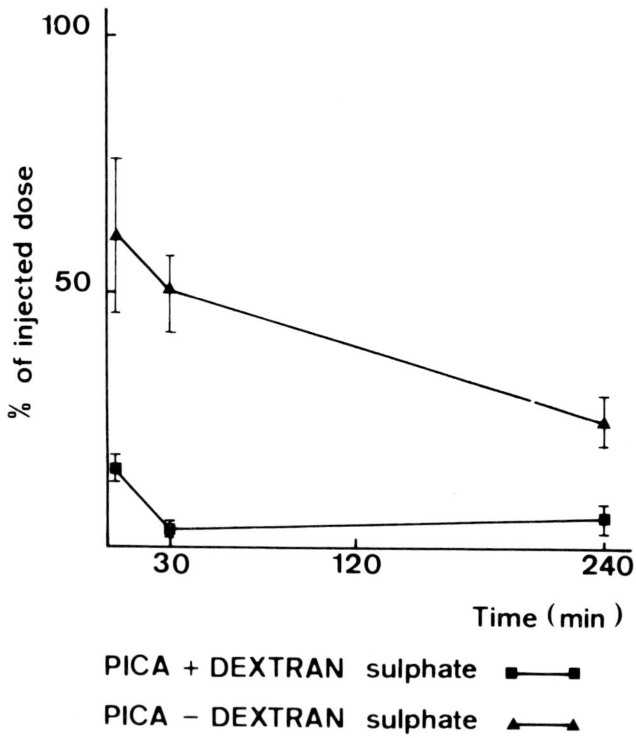

FIGURE 28. Liver radioactivity 30 min after intravenous administration of ^{14}C-PIN into mice; (■) blockaded animals, (▲) unblockaded animals.

that half of the nanoparticles remained extracellularly bound in a reversible way. The same experiment was repeated at time intervals of 5 min and 2 hr. It appeared that capture by the Kupffer cells was a rapid phenomenon (Figure 30). Indeed 5 min after injection, Kupffer cell carrier concentrations were 173-fold higher than parenchymal cell concentration, and 21.2-fold higher than endothelial cell concentration. After 30 and 120 min, Kupffer cell concentrations decreased, indicating a release of degradation products and/or particles in the vascular and extracellular spaces. On the contrary, endothelial cell concentration was the highest after 30 min, indicating a slower capture mechanism. Parenchymal cells never reached high values during the tested time intervals, showing that the radioactivity material cleared from the Kupffer cells either was carried away by the blood stream or transited via the parenchymal cells to the bile duct in a relatively fast way.

Nanoparticles used in these experiments were 215 nm diameter, which is above the size of pores in the sieve-plate of the fenestrated endothelial lining of the liver. The low uptake of particles by parenchymal cells was thought to be due to this morphological barrier. For this reason, cellular distribution of nanoparticles smaller than 80 nm was determined. Surprisingly, this drop in size did not affect the relative uptake by the different liver cells (Figure 31).

Indeed, the ratio between carrier concentration in Kupffer and parenchymal cells was respectively of 92 and 3.6 for the small particles and 101 and 4.2 for the others. The total capture by the liver for both preparations was also similar. This means that particle size has no influence on the liver distribution of the carrier and that the interaction observed between Kupffer cells and nanoparticles could be due to other factors than only liver architecture.

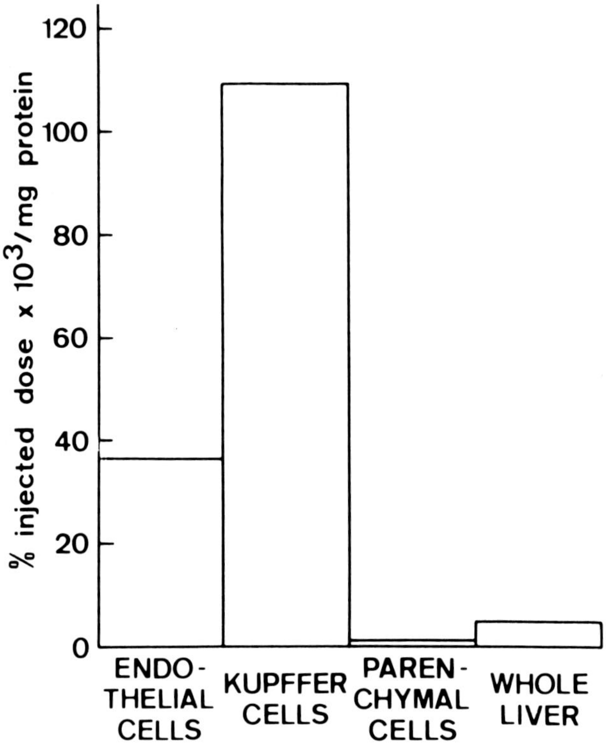

FIGURE 29. Percentage (× 10^3) of the injected dose of radioactive nanoparticles expressed per mg protein in different liver cell types and in the whole liver, 30 min after intravenous administration.

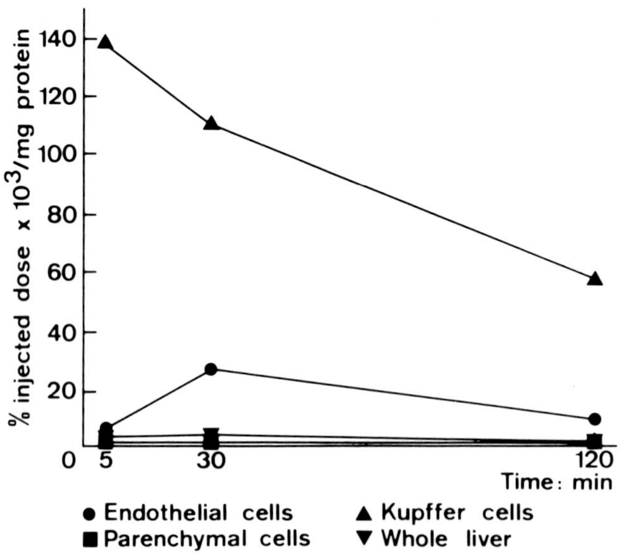

FIGURE 30. Percentage (× 10^3) of the injected dose of radioactive nanoparticles expressed per mg protein in different cell types of the liver and in total liver as function of time after intravenous administration.

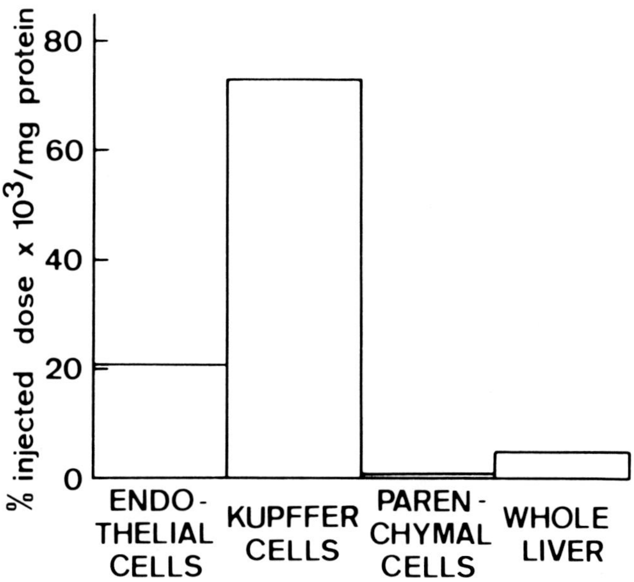

FIGURE 31. Percentage (× 10^3) of the injected dose of small size nanoparticles (0.08 μm diameter) expressed per mg protein in different cell types and in total liver, 30 min after intravenous administration.

2. Subcutaneous Administration

In the context of long acting systems it was interesting to evaluate the deportment of nanoparticles after subcutaneous injection. Furthermore, this way of administration could be of interest when hepatic overload has to be avoided. Autoradiography has been performed on mice after subcutaneous administration of ^{14}C PIN. A high radioactivity level at the site of injection was detected 30 min after the administration (Figure 32). The gut wall and the bone marrow appeared to concentrate the carrier. Curiously, the salivary glands were deeply marked. We must emphasize that no hepatic overload was detectable. The site of injection remained deeply marked for over 15 days.

3. Oral Administration

Considering the increasing demand in systems able to protect from destruction by gastrointestinal enzymes we wanted to evaluate the resorption of nanoparticles after intragastric intubation. Autoradiography was performed on mice after gastric intubation of ^{14}C PHN. As shown on Figure 33, the carrier remained exclusively in the stomach of animals 30 min after administration. After 4 hr the intestines were deeply marked, and the carrier was excreted without visible adsorption. Nevertheless, a persistent film was visible on the stomach wall. Radioactivity determined in the kidneys and the liver confirmed the lack of systemic absorption of orally administered PHN (Table 6).

B. Nanoparticles in Tumor-Bearing Animals

1. Lewis Lung Carcinoma

It was of interest to evaluate the ability of the carrier to concentrate in tumoral tissues. For the first trial, we evaluated the behavior of PIN in subcutaneously grafted Lewis lung carcinoma-bearing mice. This tumor generally gives rise to lung metastases when subcutaneously grafted. It was therefore possible to evaluate the localization of the carrier in the subcutaneous primary tumor and in the pulmonary metastatic tissue.

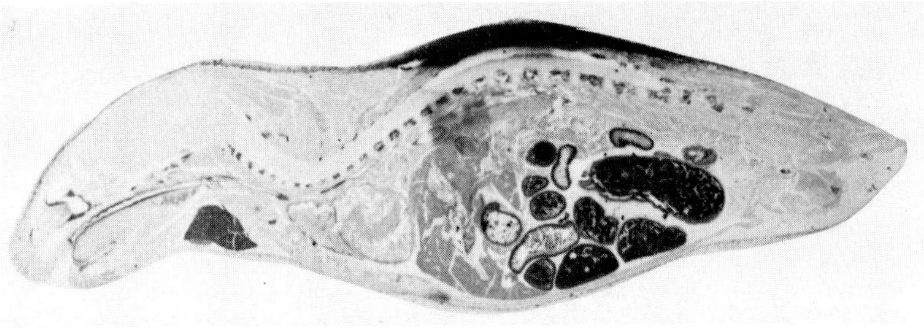

FIGURE 32. Whole body autoradiography of a mouse 30 min after subcutaneous injection of ^{14}C-labeled PIN. (From Grislain, L.et al., *Int. J. Pharm.*, 15, 335, 1983. With permission.)

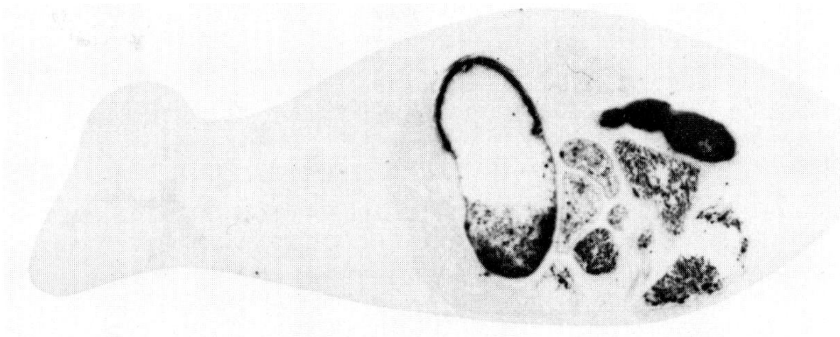

FIGURE 33. Whole body autoradiography of a mouse 4 hr after oral administration of ^{14}C-labeled PHN.

Table 6
HEPATIC AND RENAL CONCENTRATION AFTER ORAL ADMINISTRATION OF ^{14}C-POLYHEXYLCYANOACRYLATE NANOPARTICLES

Time	Kidney (% injected dose)	Liver (% injected dose)
30 min	1.3	0.6
2 hr	0.9	0.7
4 hr	0.9	0.8

From 5 min until 4 hr after the injection, progressive accumulation occurred in tumoral tissue and reached, after 4 hr a level as high as for salivary glands (Figure 34b). Furthermore, the tumor site obviously appeared more marked than the normal underlying tissue after a time interval where the nanoparticles were cleared from the blood stream. This observation confirms the localization of the carrier inside the carcinoma and implies a possible passage of the submicroscopic particles through the vascular endothelium. Furthermore, we observed a higher level of radioactivity in the lung tissue of tumor-bearing mice (Figure 34b), which did not exist in healthy animals 4 hr after injection (Figure 34a).

These observations were quantitatively confirmed by measuring the radioactivity in the lung and in the liver. In the case of tumor-bearing animals, the pulmonary radioactivity was sixfold higher than for healthy mice (Table 7). This increase in pulmonary radioactivity could be correlated to the reduction in hepatic capture.

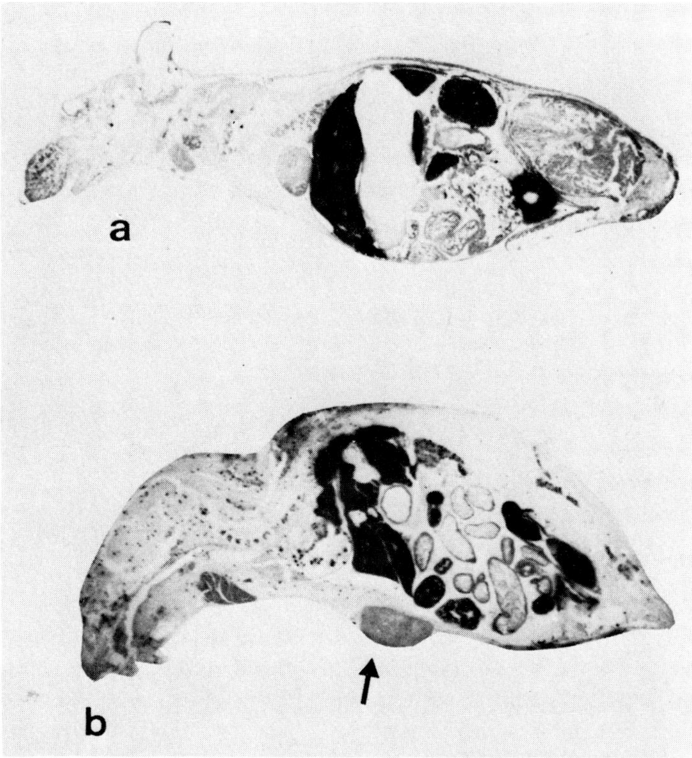

FIGURE 34. Whole body autoradiography of (b) subcutaneous Lewis Lung carcinoma (→) bearing mouse and (a) healthy mouse 4 hr after intravenous administration of ^{14}C-labeled nanoparticles. (From Grislain, L. et al., *Int. J. Pharm.*, 15, 335, 1983. With permission.)

Table 7
LUNG AND LIVER CONCENTRATION OF RADIOACTIVITY 4 HR AFTER INTRAVENOUS ADMINISTRATION OF ^{14}C NANOPARTICLES (2.5 μCi) TO HEALTHY (A) AND TUMOR BEARING MICE (B)

	Lung	Liver	Lung/liver
A	10.7 ± 1.9	26.4 ± 5.0	0.41 ± 0.13
B	67.9 ± 10.5	12.9 ± 1.0	5.26 ± 1.11

Note: The values represent the mean and the standard deviation of 7 animals in each group.

The most challenging problem in the development of a colloidal carrier results from the need to concentrate to sites different from the liver and the spleen. Our results demonstrate the possibility of such a diffusion of PIN into a neoplastic tissue.

2. Colon Adenocarcinoma

After the demonstration of the PIN capture inside primary tumoral and metastatic tissues on Lewis lung carcinoma-bearing mice, we wondered if this peculiar behavior could be repeated in other animal species and in other kinds of tumors. For this purpose, ^{14}C PIN were administered intravenously to subcutaneously grafted S447 carcinoma-bearing rats. At various time intervals after the administration, autoradiograms and quantitative determinations of the carrier concentrations were performed in several organs. As seen in Figures 35 and 36, the tumoral tissue concentrated the carrier 4 hr after the injection of the nanoparticles in the same way as observed for the Lewis lung carcinoma. As for mice, hepatic uptake was predominant.

C. Comparison Between Free and Carried Drugs

Having considered the body distribution of the carrier, the fate of targeted drugs was studied.

1. Doxorubicin (DOX)

DOX, an anthracycline antibiotic is currently used in the treatment of neoplastic diseases. However, severe side effects involving the heart (acute and chronic cardiomyopathy), the bone marrow, and the intestine frequently restrict its clinical use in medicine. We demonstrated, that DOX-loaded nanoparticles are less toxic than the free drug. This reduced toxicity could be attributed to a reduction of the capture in tissue responsible for the main toxic effect. In order to assess this probability, the tissue distribution of DOX was compared to the one of the free drug with a special emphasis on myocardium. For the first determination, 3 hr after the last of three daily injections of 10 mg/kg body weight of either free or bound DOX, mice were sacrificed and processed for whole body autofluorography. After freeze drying, cardiac sections were examined under a fluorescent microscope. DOX was easily recognized by its orange red fluorescence. Figure 37 shows intensive cardiac accumulation of the free drug. On the contrary, no fluorescence was found in the myocardium after administration of the same dose of DOX bound to nanoparticles (Figure 38).

These observations have been quantitatively confirmed; after injection of 7 mg/kg body weight of free or bound DOX to mice, plasmatic and cardiac drug levels were determined using an HPLC method. Figure 39 shows that the binding of the drug to PIN slows the drug elimination from the plasma compartment. Furthermore, the cardiac level of the adsorbed drug is drastically reduced in comparison with the free drug (Figure 40). The decrease in the cardiac uptake could be attributed to the low endocytotic activity of cardiac cells which provided a sufficiently high stability of the carrier drug complex.

2. Dactinomycin (DACT)

This study compares the tissue distribution of free and bound DACT to polymethyl-(PMN), polyethyl-(PEN), and polybutylcyanoacrylate nanoparticles (PBN) after intravenous injection to rats. For all preparations, efficiency of drug sorption on nanoparticles was determined (Table 8). For a DACT concentration of 55 $\mu g/m\ell$ in the polymerization medium, 93, 87 and 65% of the drug was adsorbed on PMN, PEN, and PBN.

Figure 41 shows the concentration of ^{3}H-DACT found in blood, spleen, small intestine, muscle, kidney, liver, and lung samples after administration of either free, PMN, or PEN adsorbed drug. Each value represents the mean of six animals. Significant increases in ^{3}H-DACT uptake were seen for the small intestine and lungs in PMN-injected rats 30 min after administration. Three hours after administration, ^{3}H-DACT concentrations decreased in both tissues. Nevertheless, the intestinal radioactivity levels remained higher than for the free drug. No significant difference was found 24 hr after injection. Similar conclusions were drawn for PEN concerning the intestine and

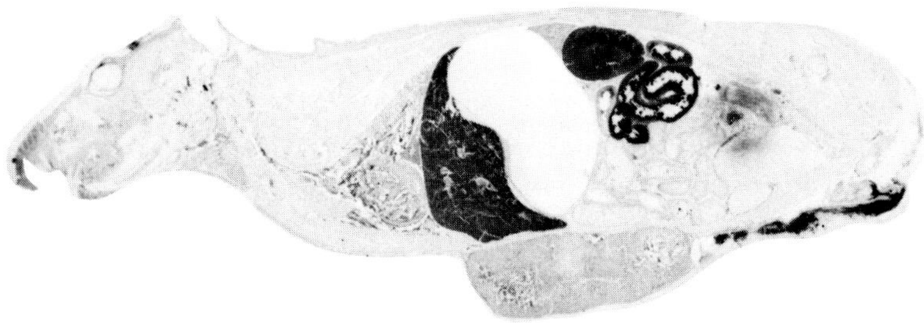

FIGURE 35. Whole body autoradiography of subcutaneous S 447 carcinoma-bearing rat, 4 hr after intravenous administration of ^{14}C-labeled nanoparticles.

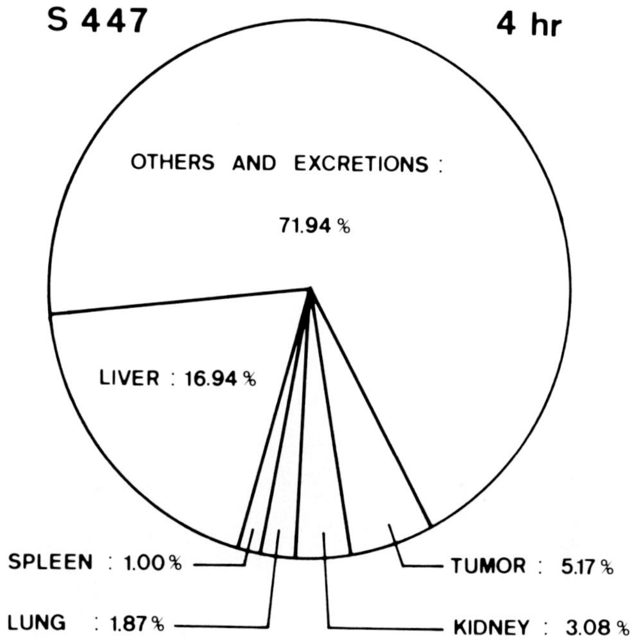

FIGURE 36. Diagram of the body distribution of ^{14}C-labeled PIN 4 hr after intravenous administration to subcutaneous S 447 carcinoma-bearing rat.

the lung. Only a slight, but significant increase in the hepatic uptake was detected for PMN and PEN adsorbed DACT in comparison with the free drug.

However, Figure 42 shows that in the PBN ^{3}H-DACT-treated rats, the concentration of ^{3}H-DACT in the liver and the spleen was importantly increased as compared to the free form for a time interval ranging from 1 to 24 hr. Unlike the free drug injected rats, PBN-treated animals showed lower concentrations of the cytostatic in the small intestine and the kidneys from 1 to 3 hr after injection.

Statistical analysis of these results indicated that PBN significantly modify the distribution pattern of the drug. Tissue concentrations observed were 64-fold higher in the liver, 44-fold higher in the spleen, and 4.7-fold higher in the lungs for the PBN associated drug in comparison with the free form. Determinations of the cumulative urinary excretion of free and adsorbed DACT showed a decrease in the urinary clearance of the drug when bound to the carrier (Figure 43). These results provide evidence that nanoparticles can be used to modify the tissue distribution of DACT. It is noteworthy that the PBN greatly increases the uptake of the drug by the tissues rich in

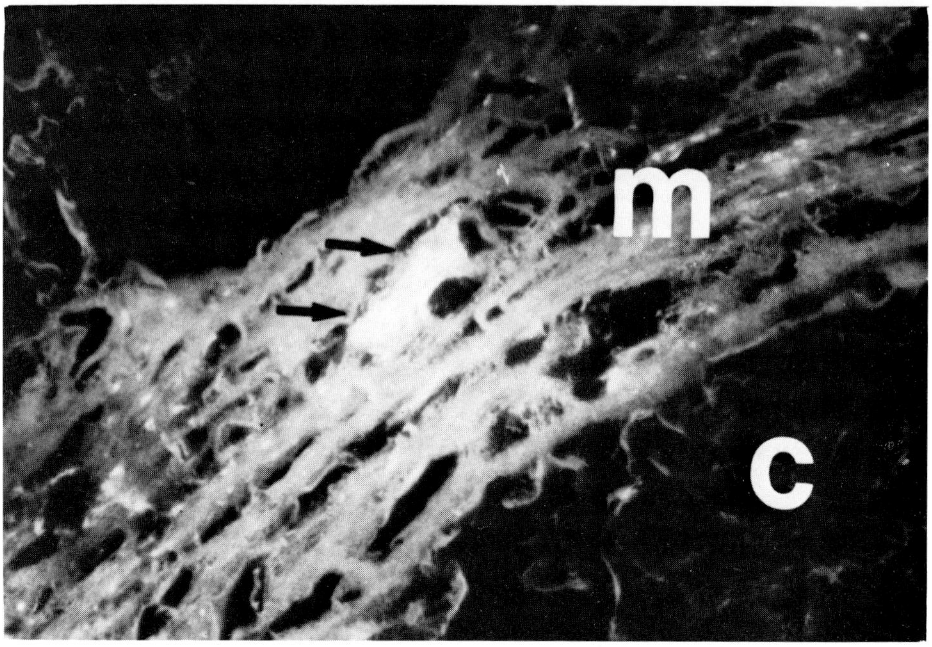

FIGURE 37. Detail of a whole body fluorogram showing the heart of a mouse sacrificed 3 hr after the last of three daily injections of 10 mg/kg of free doxorubicin. There is a high accumulation of doxorubicin in the muscular fibers of the heart (→) (m = muscular myocardic fibers; c = myocardic cavity). (From Couvreur, P. et al., *J. Pharm. Sci.,* 71, 790, 1982. With permission.)

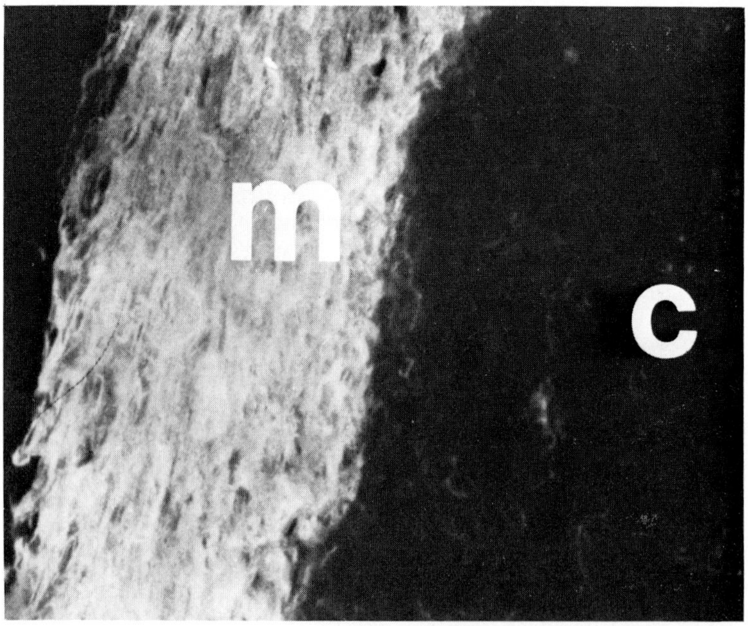

FIGURE 38. Whole body fluorogram showing the heart of a mouse sacrificed 3 hr after the last of three injections of 10 mg/kg of doxorubicin bound to nanoparticles. In comparison with Figure 37, no fluorescence appears in the myocardium (m = muscular myocardic fibers; c = myocardic cavity). (From Couvreur, P. et al., *J. Pharm. Sci.,* 71, 790, 1982. With permission.)

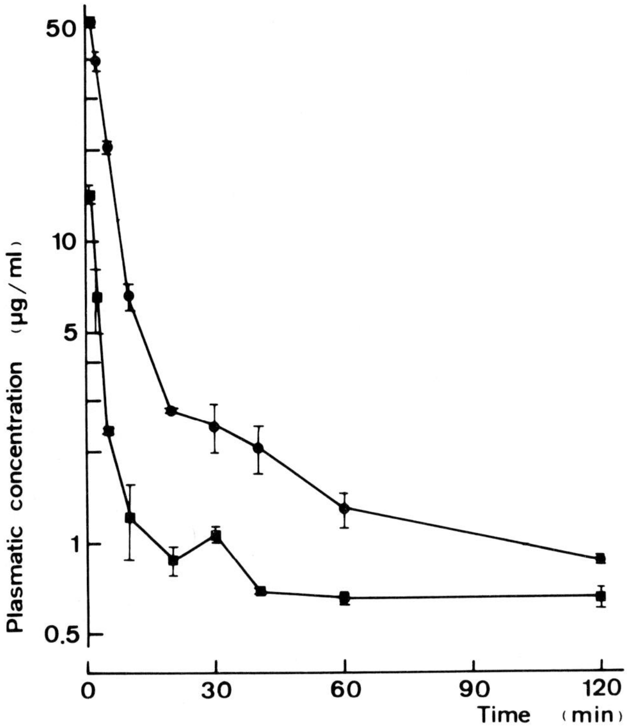

FIGURE 39. Plasmatic concentrations of free and nanoparticle-bound doxorubicin after intravenous administration into mice.

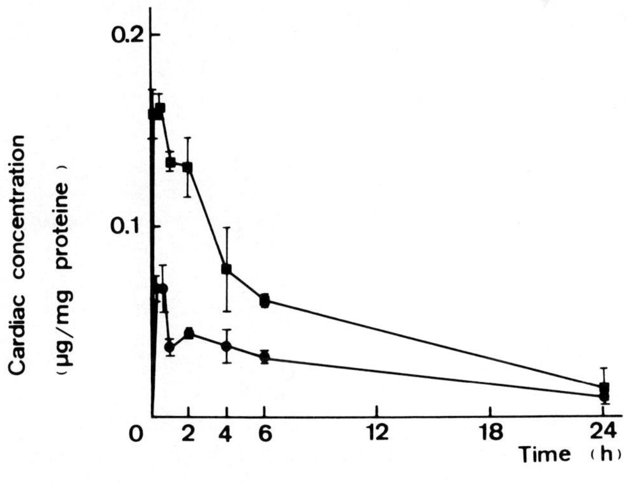

FIGURE 40. Cardiac concentration of free and nanoparticle-bound doxorubicin after intravenous administration into mice.

reticuloendothelial cells. This study also stresses the modification of the body distribution of ^{3}H-DACT when this drug is adsorbed on nanoparticles of various chemical composition.

Table 8
EFFICIENCY OF DACT DRUG SORPTION ON NANOPARTICLES OF POLYMETHYL-(PMN), POLYETHYL-(PEN), AND POLYBUTYL-(PBN) CYANOACRYLATE.

Monomer	Sorption %
PMN	92.8
PEN	86.5
PBN	65.0

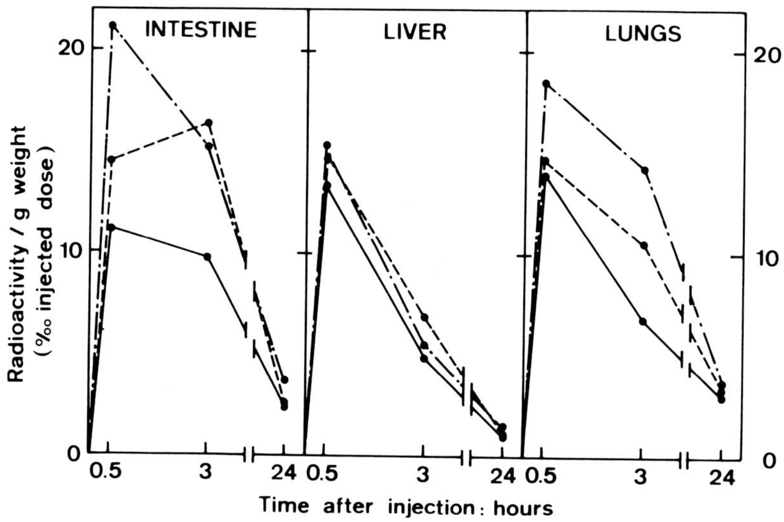

FIGURE 41. Tissue concentration in the intestine, the liver, and the lung of rats after intravenous administration of ³H-DACT. A single intravenous injection of PMN (— ·), PEN (- -) or free DACT (—) was given to each rat. Each point is the mean value from at least six animals. (From Couvreur, P. et al., *J. Pharm. Sci.*, 69, 199, 1980. With permission.)

3. *Vinblastine*

Similar experiments were carried out with ^{3}H-vinblastine (VLB). Figure 44 shows the ^{3}H-VLB tissue concentrations after injection of the free or PEN adsorbed drug. Concentrations of ^{3}H-VLB found in tissues after administration of the adsorbed drug were generally higher than those obtained with the free form. This result was particularly remarkable 30 min after the injection in tissues holding reticuloendothelial macrophages such as the liver, the spleen, and the lungs. Thus, 30 min after injection, the spleen accumulated 21-fold more nanoparticles bound ^{3}H-VBL than the free drug. Likewise, the drug concentrations observed with PEN were 1.7-fold higher in the liver, 4-fold higher in the lungs, and 2.3-fold higher in the kidneys than for free VLB. More surprising was the higher concentration of the bound ^{3}H-VLB in the muscles as compared to the free form. Differences in the tissue drug concentrations between the two formulations were generally more marked at a short time after injection.

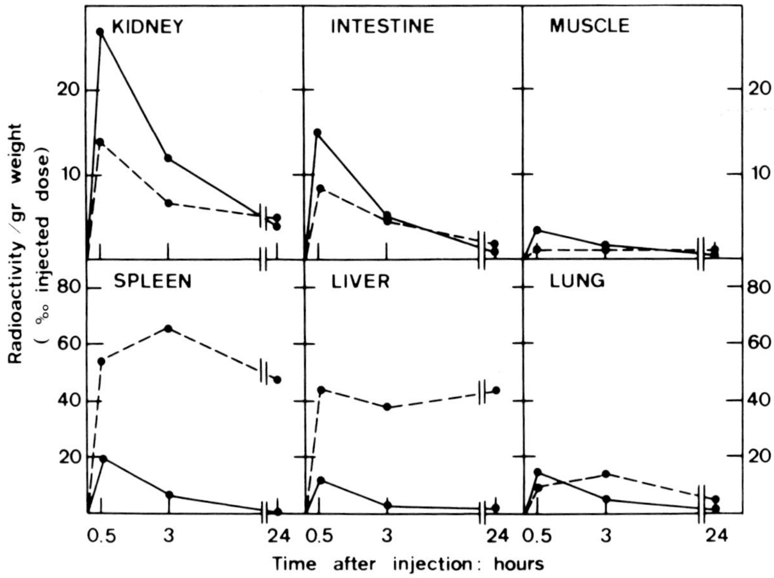

FIGURE 42. Tissue concentration of ^{3}H-DACT after a single injection into rats of free (—) or bound (---) drug on PBN. Each point represents the mean result from at least 6 animals. (From Kante, B. et al., *Int. J. Pharm.*, 7, 45, 1980. With permission.)

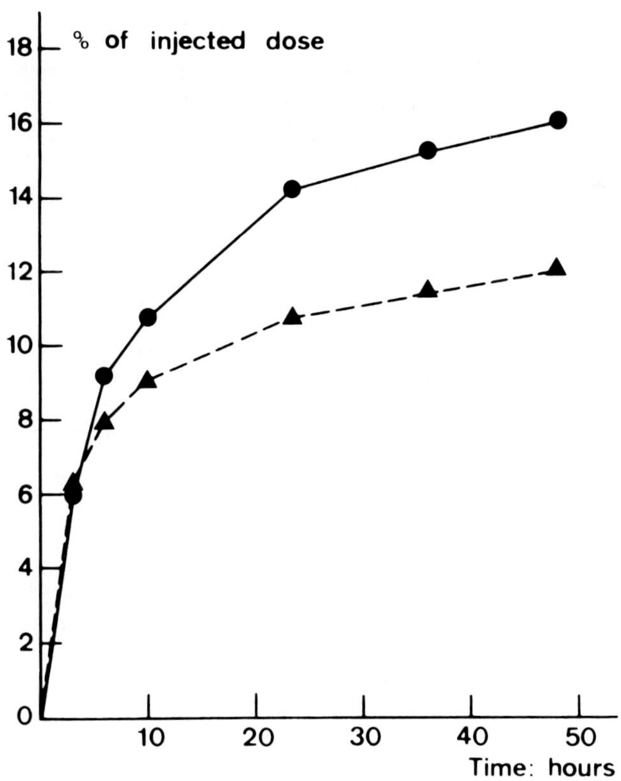

FIGURE 43. Cumulative urinary excretion of ^{3}H-DACT after a single injection into rats of free • or bound ▲ drug on PBN. Each point represents the mean value from at least 9 animals. (From Kante, B. et al., *Int. J. Pharm.*, 7, 45, 1980. With permission.)

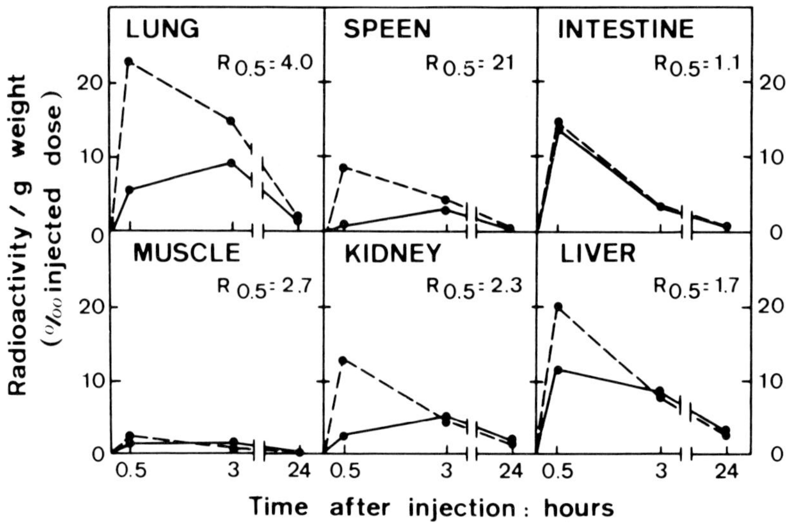

FIGURE 44. Tissue concentration (blood, spleen, intestine, muscle, kidney, liver, and lung) of free (—) and PEN (- -) ^{3}H vinblastine forms at various times after intravenous administration. Each point is the mean value from at least 6 animals. (From Couvreur, P. et al., *J. Pharm. Sci.*, 69, 199, 1980. With permission.)

V. TOXICITY OF POLYALKYLCYANOACRYLATE NANOPARTICLES

Since the discovery by Coover et al.[28] of the usefulness of alkylcyanoacrylate adhesives in bonding living tissues together, many studies have been conducted to ascertain tissue responses to these materials. In short, the methyl- and ethylcyanoacrylate homologues have been found to be relatively toxic, while the toxicity of the higher homologues decreased as the alkyl chain length increased.[29-30] Polyalkylcyanoacrylate tissue adhesives and hemostatic agents have proven useful in experimental and clinical conditions in surgery and dentistry for about 15 years.[31-32] Alkylcyanoacrylates were therefore deemed to be acceptable candidate materials for preparing polymeric nanoparticles to be used in human medicine as a drug carrier. This consideration, however, did not obviate the necessity of studying the toxicity of these particles. Up to now, in this matter, we have been trying to work out several important points rather than achieve a systematic investigation. We will report here, first, on the toxicity of the unloaded (drug-free) polyalkylcyanoacrylate nanoparticles, and second, on the possibility of lowering the inherent toxic effects of the antineoplastic agent doxorubicin by adsorbing it onto these particles.

A. Unloaded Nanoparticles

1. Mutagenicity Tests

Mutagenicity tests were performed on both intact and degraded PMN and PBN either with or without metabolic activation[33] using the *Salmonella typhimurium* method (Ames Test[34]). Briefly, PMN were prepared by adding the monomer in an aqueous solution of HCl that contained 0.5% polysorbate 20. After polymerization, the suspension was buffered to pH 7 using NaOH and a phosphate buffer, then brought to isotonicity by NaCl. No filtration was carried out before testing. PBN were prepared in the same way, except, first, that Pluronic L-63 was used in place of polysorbate (0.8% and, second, that Dextran 70 was also added to the HCl solution (0.2%). "Degraded nanoparticles" refers to a one volume suspension of nanoparticles to which two volumes of 1 *M* NaOH were added, the resulting limpid solution of degradation products being neutralized, 24 hr later, with 1 *M* HCl.

Table 9 shows the results of a preliminary assay of the bacterial survival of the *S. typhimurium* strains after incubation with nanoparticles on nutrient agar plates seeded with 2 to 7 × 10^7 viable cells. One can see that PMN induced a marked toxic effect when they were assayed at a dose of 1500 μg per plate while PBN were far less toxic at the same dose. Both forms exerted a slight inhibitory effect at the 300-μg dose. No toxicity was observed at the 150-μg dose, which was thus taken to be the maximum noninhibitory level for the two types of polymers.

The results of the subsequent mutagenicity tests are presented in Tables 10A to 10D. One can see that neither the nanoparticles (both types) nor their degradation products exerted a mutagenic effect. Indeed, the number of induced revertants as compared with spontaneous revertants was not significantly increased in any test. Furthermore, no dose-effect response was apparent in these assays.

In any mutagenicity assay, the observed mutations are related to only a very small part of the genome. Therefore, a negative Ames test does not necessarily mean that the product assayed could not be mutagenic in humans.[35] However, the Ames method has proven reliable in detecting carcinogens.[36-37] Moreover, cyanoacrylate polymers have been used in surgery for a long time, and no carcinogenic action has been observed with these products.[38-39]

2. Toxicity Towards Cells In Vitro

It has been emphasized from the outset that PAN are most likely to be phagocytosable particles.[5,40] Though this hypothesis has not been strictly proved yet, all experimental evidence has thus far supported it.[15,41] This being the case, scanning electron microscopy was utilized to assess whether morphological cell changes may arise from PAN ingestion.[33]

Macrophages, the so-called "professional phagocytes", were used in this study. They were obtained from the peritoneal cavity of C57BL mice and incubated at 37°C using standard harvesting and culturing procedures. Two nanoparticle formulations were tested: PIN, a presumably nontoxic preparation, and PMN, a form derived from an admittedly cytotoxic homologue. PIN were prepared in the same way as PBN, as described in the above paragraph wherein a short account of the PMN preparation can also be found. The nanoparticles were added at a 1% concentration in the culture medium. After 1 hr of incubation, nonadhering nanoparticles were washed off with phosphate buffer solution medium. The macrophages were then processed through fixation in glutaraldehyde and OsO_4, freezing in liquid nitrogen, freeze-drying and, finally, coating with a thin gold layer.

Scanning electron micrographs revealed marked morphological modifications after incubation of the macrophages in the presence of PMN while no signs of toxicity were observed after incubation with PIN. As a matter of fact, most of the PMN-treated macrophages showed a completely perforated cell membrane (Figure 45) while the morphology of the PIN-treated cells appeared unchanged (Figure 46) in comparison to untreated cells.

As a complementary test, the cytotoxicity of PBN and their polymerization medium was investigated in a more quantitative way using cultured hepatocytes.[33]

Hepatocytes were isolated using an in situ enzymatic perfusion method of a rat liver. They were then incubated at 37°C and PBN suspension was added in order to obtain final concentrations of 0.5 or 1% (v/v), which corresponded respectively to 7.5 or 15 mg of polymer per 100 mℓ of the incubation medium. The polymerization medium of the particles was also tested using the same concentrations (0.5 and 1%). For the purpose of assessing the viability of the incubated cells, membrane integrity was regularly controlled using both the 0.36% erythrosin B exclusion test and the leakage of lactate dehydrogenase, a typically cytosolic enzyme.

Table 9
BACTERIAL TOXICITY TEST

Salmonella typhimurium strains	Bacterial survival[a] × 10^7/plate						
		Methyl product			Butyl product		
	Controls	150 μg	300 μg	1500 μg	150 μg	300 μg	1500 μg
TA 1530	129	116	112	3	150	140	135
TA 1535	101	84	74	0	111	75	67
TA 1538	90	105	70	0	107	78	68
TA 100	91	81	65	5	109	73	69
TA 98	83	77	68	3	104	73	72

[a] Number of survivals after incubation of the strains with polymethyl- and polybutylcyanoacrylate nanoparticles on nutrient agar plates.

Table 10A
AMES MUTAGENICITY TEST WITH NANOPARTICLES WITHOUT METABOLIC ACTIVATION

Salmonella typhimurium strains	Histidine revertants[a]/plate						
		Methyl product			Butyl product		
	Spontaneous	30 μg	60 μg	150 μg	30 μg	60 μg	150 μg
TA 1530	23	19	13	19	17	17	12
TA 1535	6	10	9	12	9	14	12
TA 1538	25	14	37	15	17	9	16
TA 100	122	171	151	127	113	158	160
TA 98	44	10	14	22	23	24	23

[a] Number of revertants after incubation of the strains with polymethyl- and polybutylcyanoacrylate nanoparticles.

Table 10B
AMES MUTAGENICITY TEST INVOLVING METABOLIC ACTIVATION WITH NANOPARTICLES

Salmonella typhimurium strains	Histidine revertants[a]/plate						
		Methyl product			Butyl product		
	Spontaneous	15 μg	30 μg	150 μg	15 μg	30 μg	150 μg
TA 1530	9	8	9	5	5	11	9
TA 1535	11	7	8	10	7	9	11
TA 1538	15	8	11	8	9	11	9
TA 100	180	125	135	123	149	174	169
TA 98	14	12	13	10	13		19

[a] Number of revertants after metabolic activation and incubation with polymethyl- and polybutylcyanoacrylate nanoparticles.

Table 10C
AMES MUTAGENICITY TEST WITH THE DEGRADATION PRODUCTS OF NANOPARTICLES WITHOUT METABOLIC ACTIVATION

Salmonella typhimurium strains	Spontaneous	Histidine revertants[a]/plate: Methyl product 15 μg	30 μg	150 μg	Butyl product 15 μg	30 μg	150 μg
TA 1530	14	22	14	16	17	17	20
TA 1535	15	10	7	12	11	12	11
TA 1538	13	12	7	8	9	5	20
TA 100	145	122	137	131	168	141	138
TA 98	13	24	22	16	10	19	28

[a] Number of revertants after incubation of the strains with polymethyl- and polybutylcyanoacrylate nanoparticle degradation products.

Table 10D
AMES MUTAGENICITY TEST INVOLVING METABOLIC ACTIVATION WITH THE DEGRADATION PRODUCTS OF NANOPARTICLES

Salmonella typhimurium strains	Spontaneous	Histidine revertants[a]/plate: Methyl product 15 μg	30 μg	150 μg	Butyl product 15 μg	30 μg	150 μg
TA 1530	10	11	17	9	9	11	16
TA 1535	10	12	12	13	15	7	12
TA 1538	28	36	—	29	27	—	29
TA 100	178	137	—	120	170	—	133
TA 98	46	47	39	40	39	35	44

[a] Number of revertants after metabolic activation and incubation with polymethyl- and polybutylcyanoacrylate nanoparticle degradation products.

At the 0.5% dose, neither PBN nor their polymerization media modified the cellular integrity of the hepatocytes. Indeed, as shown in Figure 47, both the lactate dehydrogenase leakage and the dye exclusion capacity of the treated cells remained similar to those of the control cells. At the 1%-dose however, PBN greatly affected the integrity of the hepatocytes while the polymerization medium exerted no effect (Figure 48). Thus, a cytotoxic effect appeared between 0.5 and 1%, and seemed to proceed form the nanoparticles themselves rather than from the surfactive agent employed in their preparation. However, the 1% dose corresponded to a relatively high particles-to-hepatocytes ratio ($\sim 10^5$:1).

These preliminary in vitro experiments did not reveal any severe toxicity that could restrict the use of polyalkylcyanoacrylate nanoparticles as a drug carrier. Indeed, these particles induce cellular damage only at a relatively high concentration in the culture medium (1%). This effect is probably due to the presence of degradation products in the cytoplasm following phagocytosis of the particles. This consideration could explain the lower toxicity of the isobutyl homologue, which degrades more slowly than the methyl homologue.

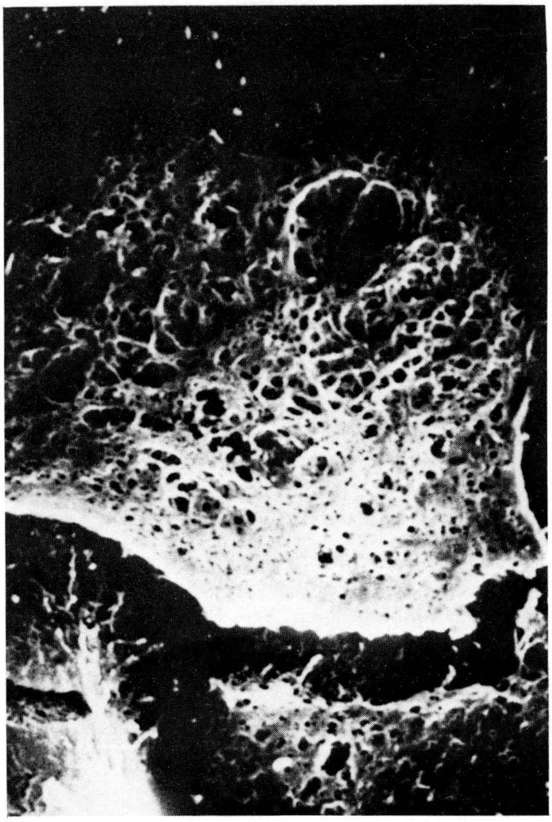

FIGURE 45. Scanning electron microscopic appearance of a mouse peritoneal macrophage incubated 1 hr with PMN at a 1% concentration in the culture medium. From Kante, B. et al., *J. Pharm. Sci.*, 71, 786, 1982. With permission.)

3. Acute Toxicity in Mice

As an estimation of the acute toxicity of nanoparticles, groups of 10 NMRI male mice (20 to 25 g) were used to determine the 50% lethal dose (LD_{50}) of PBN, PIN, and PHN by single rapid intravenous injection.[33,42]

PBN and PIN were prepared as described in the above paragraphs, except that the suspensions were passed through a sintered glass filter (pore size 9 to 15 μm) as an additional step; this resulted in a final concentration of 9.2 mg of polymer per milliliter of either PBN or PIN suspension. In both cases, six different doses ranging from 12.5 to 40.0 mℓ of suspension per kilogram of body weight (i.e., from 115 to 368 mg of polymer per kilogram of body weight) were tested. Another six groups of mice were used to test the polymerization medium of the particles in a comparative fashion.

Based on the cumulative mortality of the mice as registered within the next 24-hr period after the injection (Figure 49), the LD_{50} was estimated at 196 and 230 mg/kg for PIN and PBN respectively. However, presumably due to the surfactive agent it contained, the polymerization medium itself was not lacking toxicity (LD_{50} = 33.4 mℓ/kg). From this observation, it was concluded that a significant lowering of the acute toxicity of nanoparticles probably could be achieved by removing the surfactant from the suspension while the same held true with respect to the possible remaining monomers.

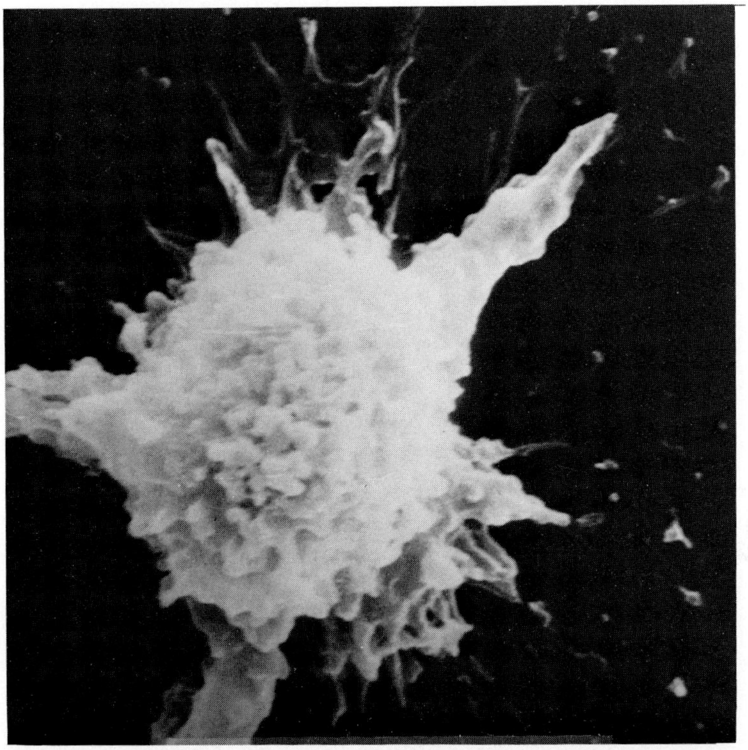

FIGURE 46. Scanning electron microscopic appearance of a mouse peritoneal macrophage incubated 1 hr with PIN at a 1% concentration in the culture medium. (From Kante, B. et al., *J. Pharm. Sci.*, 71, 786, 1982. With permission.)

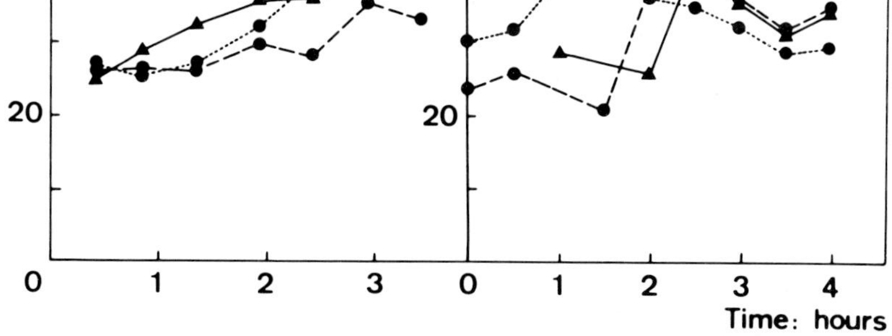

FIGURE 47. Hepatocyte mortality after incubation with 0.5% of PBN (.....) and 0.5% of nanoparticle polymerization medium (----) compared to control cells (——). (A) Dye exclusion test (erythrosin B); (B) lactate dehydrogenase leakage. (From Kante, B. et al., *J. Pharm. Sci.*, 71, 786, 1982. With permission.)

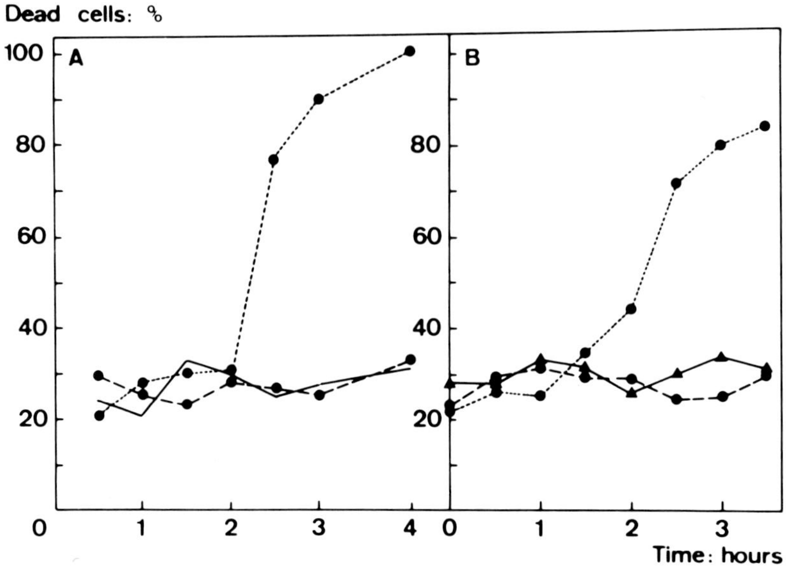

FIGURE 48. Hepatocyte mortality after incubation with 1% of PBN (.....) and 1% of nanoparticle polymerization medium (----), compared to control cells (——). (A) Dye exclusion test (erythrosin B); (B) lactate dehydrogenase leakage. (From Kante, B. et al., *J. Pharm. Sci.*, 71, 786, 1982. With permission.)

In the case of the PHN preparation, the hexylic monomer was added in an aqueous solution of 0.05 *M* H_3PO_4 containing 1% Dextran 70. After polymerization, the particles were collected by allowing the suspension to centrifuge at 21,000 rpm for 2 hr. They were then redispersed in physiological glucose, at a concentration of 50 mg of polymer per milliliter of suspension. From this preparation, seven different dose levels ranging from 10 to 13 mℓ/kg of body weight (i.e., from 500 to 650 mg of polymer per kilogram of body weight) were tested using groups of ten mice. The mortality rate was recorded for the 14 days following the injection. As shown in Figure 50, the LD_{50} of PHN was estimated at 585 mg/kg while the highest nonlethal dose was as high as 525 mg/kg under these experimental conditions. Thus, PHN were found to be significantly less toxic than PBN or PIN upon acute intravenous administration into mice, a result that was in agreement with the known low histotoxicity of the heavier alkylcyanoacrylate homologues. However, the tested PHN suspension was free of surfactive agent whereas the PBN and PIN preparations contained Pluronic L-63, and this may have accounted, to a certain extent, for the fact that PHN were far less toxic in these assays. Along the same line, the fivefold higher polymer concentration of the PHN suspension (50 mg/mℓ) allowed large quantities of nanoparticles to be administered into the mice while using small injection volumes (0.2 to 0.3 mℓ per mouse), whereas greater volumes had to be used in the PBN and PIN tests (up to 1.0 mℓ per mouse); it is uncertain, however, whether such a change in the experimental conditions had an effect upon the results observed. In any case, all three types of nanoparticles were considered worthy material for further experimentation in view of their moderate acute toxicity in mice.

It should also be stated that PMN, PEN, and PBN were tested for histotoxicity by subcutaneous administration of 0.2 mℓ of nanoparticle suspension (10 mg of polymer per milliliter) into the back of NMRI mice weighing 20 to 25 g. Twenty-four hours after the injection, the animals were sacrificed by CO_2 asphyxiation; the skin was turned inside out and tissues were examined macroscopically according to a previously described method.[43] Neither necrosis nor signs of tissue irritation were noted in any test.[44]

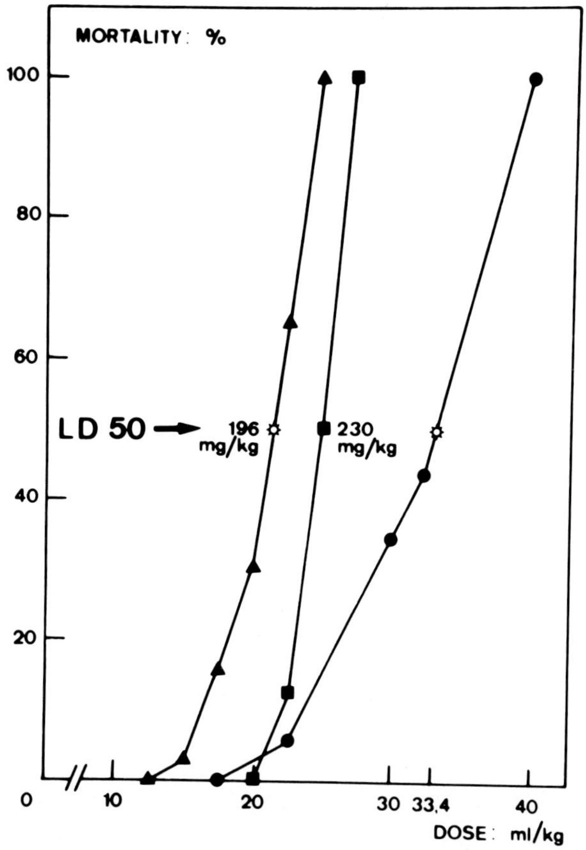

FIGURE 49. Percent of dead mice 24 hr postintravenous administration of various dose volumes of nanoparticle suspensions. (▲) PIN; (■) PBN; (•) nanoparticle polymerization medium. (From Kante, B. et al., *J. Pharm. Sci.*, 71, 786, 1982. With permission.)

4. Subacute Toxicity in Rats

Another study was carried out in order to ascertain the toxicological responses of rats to repeated intravenous injections of PIN and PHN.

Each of these formulations was given to groups of five rats in a two-injection-per week × 4-week schedule (short-term assay), and in a one-injection-per week × 16-week schedule (long-term assay), at doses of 20 mg of polymer per kilogram of body weight per injection in both the short-term and long-term assays. PIN and PHN were prepared by adding 0.07 mℓ of isobutylic or hexylic monomer to a 10 mℓ portion of a 5% glucose and 0.05 *M* phosphoric acid solution under stirring. After polymerization, the suspensions were adjusted to pH 7.0 with 0.5 *M* NaOH. The nanoparticle size as measured in a Coulter nano-sizer apparatus was equal to 185 and 145 nm for PIN and PHN respectively. The particles were suspended in their polymerization medium at such a concentration that a volume of 2.86 mℓ/kg of rat body weight per injection was delivered. They were administered by tail vein injection under ether anaesthesia. Besides the test groups, three control groups of five rats each were scheduled in both the short-term and long-term assays: (1) anaesthetized rats injected intravenously with the polymerization medium of the particles at a dose of 2.86 mℓ/kg per injection (PMED), (2) anaesthetized, noninjected rats, and (3) nonhandled rats. Wistar male rats (160 to 200 g) from a commercial source of specific pathogen-free animals were used in this study.

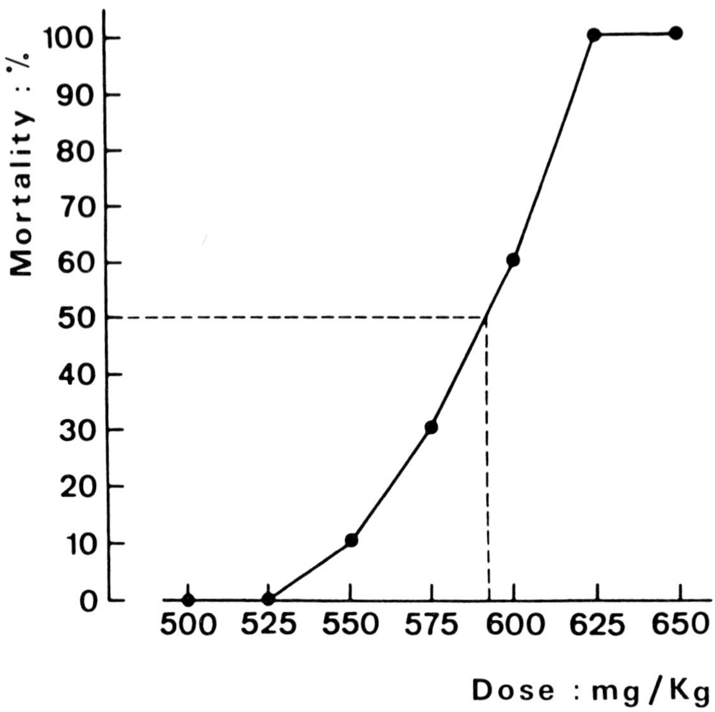

FIGURE 50. Percent of dead mice 14 days postintravenous administration of various doses of PHN.

They were kept under conventional housing conditions in plastic cages of five individuals while being provided with a commercial pelleted diet (U.A.R. A03; Villemoison-sur-Orge, France) and clean water *ad libitum*. Room temperature was maintained at 23°C with 45 to 50% relative humidity, and a regular diurnal lighting cycle of 12 hr each of light and dark was provided. The animals were weighed regularly. Two days (in the short-term assay) or 12 days (in the long-term assay) after being given the last injection of the test products, blood samples were taken from the abdominal vena cava under ether anaesthesia (after a 15-hr period of fasting), immediate necropsy examination was performed, and selected tissues were removed for light microscopic examination. The tissues were fixed in 10% neutral formaldehyde solution or in Carnoy's fluid and then processed for paraffin embedding. The paraffin slices were stained with hematoxylin-eosin or by the P.A.S. method.

With the exception of an accidental death, during anaesthesia, in the long-term PMED group, no mortality was noted during the assays. In the short-term assay, a similar increase in body weight (with respect to time) was observed (Figure 51) in all rat groups regularly subjected to ether anaesthesia, irrespective of the fact that the animals were administered either nanoparticles (PIN or PHN) or polymerization medium (PMED) or no injection. The nonhandled control group gained a little more weight than the other four. However, no significant difference was found among the five experimental groups using variance analysis (Fischer's test) of the mean body weight values registered at the end of the survey period (28th day). The same general results were observed in the long-term assay (Figure 52).

Marked morphological changes were observed in only two rats in these assays. First, in the short-term assay, a PIN-injected rat showed bilateral hydronephrosis following a partial block of lower urinary outflow. This block was most probably caused by either a yellowish concretion (found in the dilated urinary bladder), or by compression

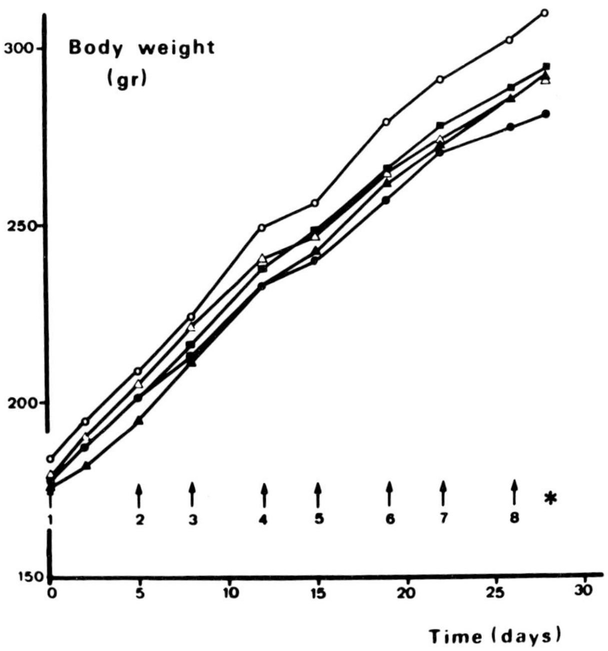

FIGURE 51. Change in body weight with respect to time in rats injected intravenously with PIN (•) and PHN (■). Dosage: 8 × 20 mg of polymer/kg of body weight (↑). Control groups: nanoparticle polymerization medium-injected rats (▲); anaesthetized, noninjected rats (△); nonhandled rats (○).

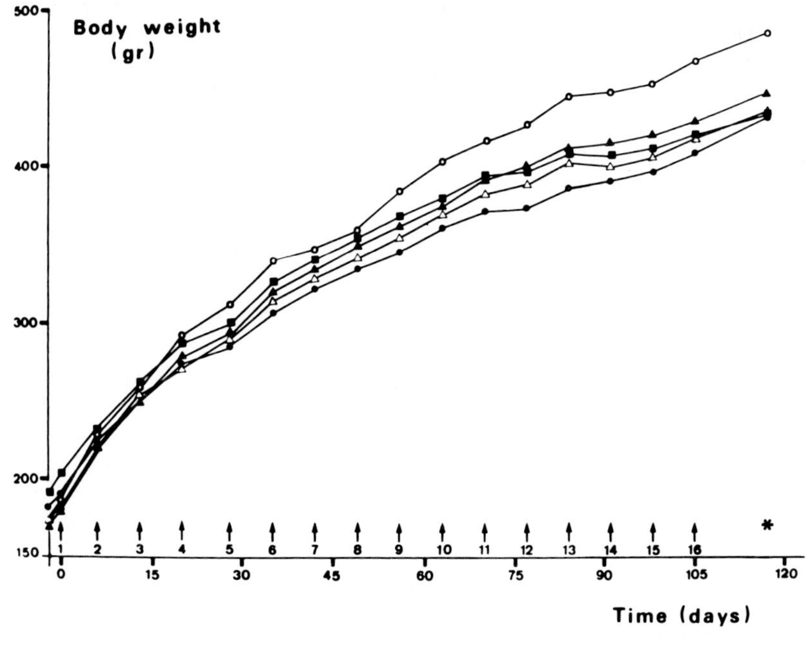

FIGURE 52. Change in body weight with respect to time in rats injected intravenously with PIN (•) and PHN (■). Dosage: 16 × 20 mg of polymer/kg of body weight (↑). Control groups: nanoparticle polymerization medium-injected rats (▲); anaesthetized, noninjected rats (△); nonhandled rats (○).

due to inflammatory-hemorrhagic intumescence of the prostate and other adjoining sexual glands (with dilated lumen of ducts). Serum creatinine was as high as 7.5 mg/dℓ in this rat. Second, in the long-term assay, a polymerization medium-injected rat exhibited gross hemorrhagic and necrotic lesion (in the stage of regeneration) in the left lateral lobe of the liver. Consistent with this change, we also found a striking serum-GPT elevation (391 IU/ℓ), a significant decrease in the RBC count, hemoglobin level and hematocrit, as well as signs of hemopoiesis stimulation in the bone marrow. Given their gross hemorrhagic component, the lesions that we observed in the pelvic region (on the first rat) and in the abdominal region (in the second rat) were most probably traumatic. It can be reasonably assumed that in both cases they resulted from inadvertant mishandling of the rats, or from a too-rapid intravenous injection.

Regarding the other rats, Tables 11 and 12A show that serum creatinine, serum γ-GT, and serum-GPT were normal in all experimental groups. The same held true for the hematologic parameters (Tables 11 and 12B), although noticeable differences in the platelet count were noted between the treated groups (PIN or PHN) and the nonhandled group in the long-term assay. The RBC count, hemoglobin level, and hematocrit were higher, generally, in the long-term assay than in the short-term assay, an observation that might be the result of the difference in rat age (24- or 11-week-old rats, respectively) at the time that blood samples were taken. The sodium and chloride serum concentrations were normal in all rat groups in both assays (Tables 11 and 12A), whereas no reliable conclusions could be drawn from the potassium determinations (due to hemolysis of a number of blood samples, which caused spurious elevations of the serum K level). We then carried out another such short-term toxicity assay with 15 PIN-injected (8 × 20 mg/kg i.v., with a 3-day interval between injections) and 15 non-injected, anaesthetized rats. Blood samples were taken from five injected and five non-injected rats at different time intervals (24 hr, 48 hr and 7 days) after the last injection (with care being taken to promptly separate serum from cells), and no significant variations in the serum potassium level were observed (Table 13A). Similarly, this experiment allowed us to confirm that repeated intravenous injections of PIN into rats had no effect on the serum sodium level (Table 13A) or on the hematologic pattern (Table 13B).

Most of the organs and tissues examined microscopically showed no changes either in the nanoparticle-treated rats or in the control rats, nor in the short-term or the long-term assays. These organs and tissues included the cerebrum and cerebellum, stomach, small and large intestine, testis and epididymis, pancreas, adrenal gland, thymus, mesenteric lymph nodes, and femoral bone marrow. No changes were observed in the liver, except a significant glycogen depletion of the hepatocytes in all rats subjected to ether narcosis (i.e., the PIN-, PHN-, and PMED-injected groups, and the anaesthetized non-injected group). Small lesions of the kidneys such as slight passive hyperemia and rare interstitial mononuclear infiltrations were noted in the short-term assay. The same kind of lesions were more distinct in the long-term assay. In the latter assay, we also observed, throughout the kidneys, some glomeruli of atrophic appearance, or glomeruli whose basal membrane structure was blurred, or whose capillaries had thickened walls in places and/or exhibited internasal adhesions; furthermore, a few nephrons showed dilated tubules containing hyaline casts. These renal lesions were spread rather evenly throughout all five rat groups of the long-term assay. It is possible to consider them as early aging-associated changes. The lungs showed small foci of emphysema, atelectasis and hyperemia, swelling of some alveolar septa, and interstitial mononuclear infiltrations; small foci of interstitial mononuclear reaction were also observed in the myocardium. None of these lesions seemed to affect any particular group specifically, but were generally less marked in the rats of the short-term assay.

Table 11
VALUES OF CERTAIN HEMATOLOGIC AND SERUM PARAMETERS IN RATS AFTER INTRAVENOUS ADMINISTRATION OF PIN, PHN, AND PMED IN A TWO-INJECTION PER WEEK × 4-WEEK SCHEDULE

	Crea (mg/dℓ)	Na (mEq/ℓ)	SGPT (IU/ℓ)	γGT (IU/ℓ)	Hb (g/dℓ)	Hct (%)	RBC ($10^6/mm^3$)
PIN	0.4—0.5	140—142	25—38	13—14	12.7—13.3	29.6—42.1	6.8—6.8
	0.42	141.0	32.7	13.7	13.00	35.80	6.84
PHN	0.4—0.9	139—142	30—53	13—15	12.0—14.3	31.3—40.1	6.7—8.0
	0.52	140.8	39.4	14.0	12.96	36.32	7.18
PMED	0.4—1.3	141—145	19—44	14—15	11.6—14.1	37.4—48.1	6.6—7.9
	0.72	142.8	30.2	14.4	13.48	44.54	7.37
Control[a]	0.3—0.9	140—143	11—41	13—15	11.4—14.7	33.2—44.6	6.0—8.0
	0.52	141.6	32.0	13.8	13.34	38.38	7.21
Control[b]	0.3—1.1	143—145	12—51	13—16	13.0—15.4	37.4—50.7	7.6—8.5
	0.82	143.6	30.2	14.4	14.40	44.90	7.86

Note: The lowest, highest, and average values of each group of 4 (in the case of PIN) or 5 rats are given.

[a] Anaesthetized noninjected rats.
[b] Nonhandled rats.

Table 12A
VALUES OF CERTAIN SERUM PARAMETERS IN RATS AFTER INTRAVENOUS ADMINISTRATION OF PIN, PHN, AND PMED IN A ONE-INJECTION PER WEEK × 16-WEEK SCHEDULE

	Crea (mg/dℓ)	Na (mEq/ℓ)	Cl (mEq/ℓ)	SGPT (IU/ℓ)
PIN	0.5—0.6	140—143	101—109	39—60
	0.52	141.6	106.2	48.4
PHN	0.5—0.7	141—143	104—108	37—66
	0.54	142.2	106.2	44.6
PMED	0.5—0.7	145—147	107—108	38—45
	0.55	146.0	107.7	40.7
Control[a]	0.5—0.5	146—147	107—110	34—61
	0.50	146.4	109.0	45.8
Control[b]	0.5—0.6	145—153	107—112	35—50
	0.54	147.0	108.8	45.0

Note: The lowest, highest, and average values of each group of 3 (in the case of PMED) or 5 rats are given.

[a] Anaesthetized noninjected rats.
[b] Nonhandled rats.

In general, in these subacute toxicity assays, PIN and PHN were found to have no significant effect on either the histological pattern of the tissues studied or on the blood parameters analyzed, or on body weight.

Table 12B
VALUES OF CERTAIN HEMATOLOGIC PARAMETERS IN RATS AFTER INTRAVENOUS ADMINISTRATION OF PIN, PHN, AND PMED IN A ONE-INJECTION PER WEEK × 16-WEEK SCHEDULE

	Hb (g/dℓ)	Hct (%)	RBC (10^6/mm^3)	WBC (10^3/mm^3)	Plts (10^3/mm^3)
PIN	14.6—15.9 15.42	44.4—49.0 46.88	8.3—9.5 8.97	6.1—7.3 6.66	516—834 724
PHN	14.4—16.1 15.32	45.9—49.1 47.50	8.7—9.4 9.19	4.6—8.7 7.16	566—792 726
PMED	14.5—15.6 14.93	45.1—48.4 46.57	8.4—8.9 8.68	5.3—8.8 7.37	606—794 668
Control[a]	13.5—15.6 14.64	44.6—49.3 47.20	8.4—9.1 8.79	5.6—7.3 6.12	546—766 623
Control[b]	14.6—16.5 15.68	46.5—50.5 48.64	8.6—9.5 9.04	6.1—8.4 6.94	492—636 547

Note: The lowest, highest, and average values of each group of 3 (in the case of PMED) or 5 rats are given.

[a] Anaesthetized noninjected rats.
[b] Nonhandled rats.

Table 13A
SODIUM AND POTASSIUM SERUM LEVELS IN RATS AFTER INTRAVENOUS ADMINISTRATION OF PIN (8 × 20 MG/KG BODY WEIGHT WITH A 3-DAY INTERVAL BETWEEN INJECTIONS), COMPARED TO (NONINJECTED) CONTROL RATS (C)

	Na (mEq/ℓ)		K (mEq/ℓ)	
	PIN	C	PIN	C
24 Hr	146.0	146.0	3.55	3.80
48 Hr	145.5	144.5	4.00	3.85
7 Days	144.5	144.5	3.50	3.85

Note: Rats were sacrificed at 3 different time intervals after the last injection. Each value represents the mean of 5 rats.

B. Doxorubicin-Loaded Nanoparticles

Doxorubicin's impressive record of activity against a broad spectrum of tumors has established it as one of the most useful drugs in current oncologic practice.[45,46] However, cardiomyopathy, myelosuppression, and gastrointestinal disturbances are three major dose-limiting complications of the drug.[47,48] Many attempts at reducing doxorubicin's toxicity have been made in recent years. One of these consists of altering the

Table 13B
VALUES OF CERTAIN HEMATOLOGIC PARAMETERS IN RATS AFTER INTRAVENOUS ADMINISTRATION OF PIN (8 × 20 MG/KG BODY WEIGHT WITH A 3-DAY INTERVAL BETWEEN INJECTIONS), COMPARED TO (NONINJECTED) CONTROL RATS (C)

	Hb (g/dℓ)		Hct (%)		RBC ($10^6/mm^3$)		WBC ($10^3/mm^3$)		Plts ($10^3/mm^3$)	
	PIN	C	PIN	C	PIN	C	PIN	C	PIN	C
24 Hr	16.1	15.9	49.0	48.2	8.11	7.97	5.8	5.7	760	892
48 Hr	16.1	16.3	49.5	50.0	8.18	7.87	5.8	5.2	649	751
7 Days	15.6	15.5	47.2	46.4	7.77	7.59	7.3	6.0	684	721

Note: Rats were sacrificed at 3 different time intervals after the last injection. Each value represents the mean of 5 rats.

in vivo distribution pattern of the drug by binding it to macromolecular carriers such as DNA[49,50] or by using it in its liposome-encapsulated form.[51] Therefore, we felt it was worth testing whether PAN could be useful in lowering the unwanted effects of doxorubicin, especially as the drug had been found to be highly adsorbed onto these particles (see Section II.C).

PIN were prepared in the presence of doxorubicin and brought to isotonicity with sodium chloride after polymerization. The polymer content of the suspension was 10 mg/mℓ while the drug was used at a concentration of 1 mg/mℓ. For the purpose of determining the fixation rate of doxorubicin to PIN, fluorometric analyses were performed on both the supernatant and the sediment after a test suspension had been allowed to centrifuge at 21,000 rpm for 2 hr. It was shown that 94% of the drug used was bound to the nanoparticles, a value that corresponded to a drug-to-polymer ratio of 94 mg/g.

The comparative toxicity study[52] of the free drug (DOX) vs. the PIN-bound drug (PIN-DOX) was conducted using groups of ten NMRI female mice (20 to 25 g) which were treated intravenously once a day for 2, 3, or 5 consecutive days.

Figure 53 shows the survival of the mice injected on 3 consecutive days. Regardless of whether doxorubicin was free or not, no deaths were observed at drug doses of 5.0 and 7.5 mg/kg/day (not shown in the Figure), whereas 15 mg/kg/day was clearly an overdose. In the middle-dosed test groups (10 and 12.5 mg/kg/day), the longer survival of the PIN-DOX-treated mice was very apparent. Other evidence of the lower overall toxicity of the bound drug form lay in the more favorable body weight evolution of the mice thus treated (Figure 54). At the 5 mg/kg/day dose, for instance, the maximal body weight loss (as compared to body weight on day zero) of the DOX group was twice as much as it was for the PIN-DOX group. In fact, the weight loss of the DOX group was significantly higher than that of the PIN-DOX group (student's *t*-test; $p > 0.975$) at all three doses tested (5, 7.5, and 10 mg/kg/day). Moreover, at the 10 mg/kg/day dose, the weight difference between the two test groups (Figure 54A) was most likely underestimated. Indeed, no dead mice were noted in the PIN-DOX group, whereas those mice which died among the DOX group were those which had lost the most weight, and they were subsequently dropped out of the weight loss computation.

When given on 2 consecutive days (instead of on 3 consecutive days), the doxorubicin doses of 12.5 and 15 mg/kg/day gave rise to better results with the bound drug than with the free drug (Figures 55 and 56). As mentioned above for the 3 × 10 mg/kg dose groups, it should be noted that the body weight loss difference between the 2 × 15 mg/kg dose groups (Figure 56A) was also most likely underestimated.

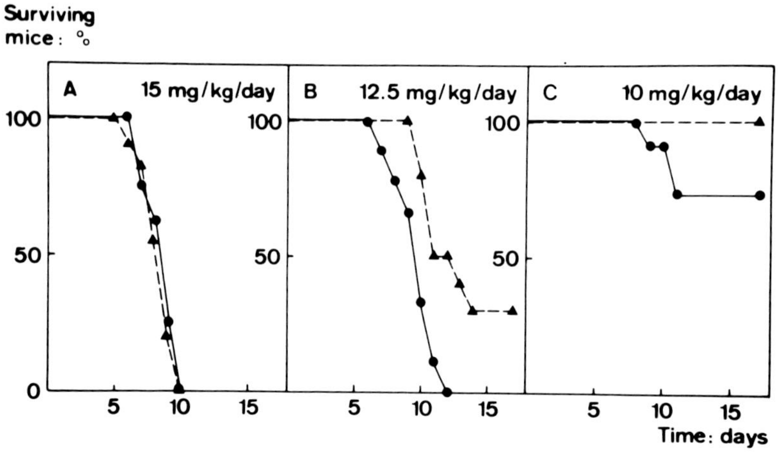

FIGURE 53. Percent of surviving mice after intravenous administration on 3 consecutive days of various doses of free DOX (•); or DOX adsorbed on PIN (▲). (From Couvreur, P. et al., *J. Pharm. Sci.*, 71, 790, 1982. With permission.)

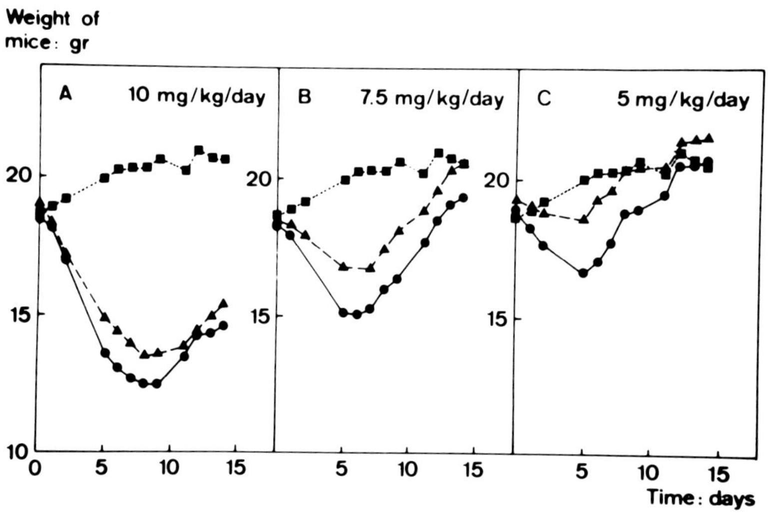

FIGURE 54. Change in body weight with respect to time in mice injected intravenously on 3 consecutive days with various doses of free DOX (•); or DOX adsorbed on PIN (▲). (■) Represents the weight of control mice. (From Couvreur, P. et al., *J. Pharm. Sci.*, 71, 790, 1982. With permission.)

Finally, in the five-injection experiments (wherein drug doses of 5.0 and 8.0 mg/kg/day were assayed), the PIN-DOX-injected mice survived longer than the DOX-injected ones (Figure 57).

As the next step in this toxicity study, we soon began comparing DOX and PIN-DOX for their effects on the weight and the histological pattern of several organs in mice. We also collected comparative data on the intestinal mucosa thickness, and on the megakaryocyte count in the femoral bone marrow. BDF1 hybrid male mice (23 to 26 g) were used. Two groups were given either DOX or PIN-DOX intravenously, at a nonlethal dose (7.5 mg of drug/per kilogram of body weight per day × 3 consecutive days, with the daily dose-volume of either DOX solution or PIN-DOX suspension being 0.25 mℓ/25 g of body weight). Polymerization medium-injected mice were used as a control group (0.25 mℓ/25 g of body weight per day i.v., × 3 consecutive days).

Table 14
COMPARISON OF THE EFFECTS OF DACT AND PMN-DACT IN CHEMOTHERAPY OF S250 SARCOMA

	Experiment schedule			Tumor size (mm²)[c]			Tumor size ratio		Mortality[f]	
Exp. no.	Dose[a]	Days of injection[b]	Day of tumor measure[b]	C[d]	PMN-DACT	DACT	PMN-DACT / DACT	t[e]	PMN-DACT	DACT
1	3.30	10 & 15	14	926	313	672	0.47	6.536	0	0
			17	1451	467	658	0.71	2.206	1	0
			21	(s)	254	1037	0.25	8.451	3	0
2a	3.30	11 & 13	17	1764	357	955	0.37	8.951	1	0
2b	3.30	12 & 14	17	1764	541	930	0.58	1.776	2	0
2c	3.30	13 & 14	17	1764	471	1854	0.25	16.611	4	2
3	3.30	12 & 14	19	1043	0	235	0.00	2.370	2	0
			24	(s)	32	684	0.05	2.739	4	0
4a	1.65	17 & 18	20	722	403	651	0.62	3.099	0	0
			24	1239	633	1012	0.63	2.858	0	0

[a] Volume (mℓ) of either PMN-DACT or DACT per kg body weight per injection; dactinomycin concentration in both PMN-DACT and DACT: 222 μg per 3.30 mℓ.

[b] The day of tumor graft was considered as day 0.

[c] Means from surviving rats.

[d] Control group.

[e] Results of student's t-test; underlined value indicates that the difference in tumor size between PMN-DACT and DACT is significant ($p = 0.975$).

[f] Total numbers of deaths recorded at the day of measure (initial groups contained six rats). No death was observed among controls. (s) Sacrificed.

From Brasseur, F., et al., *Eur. J. Cancer,* 16, 1441, 1980. With permission.

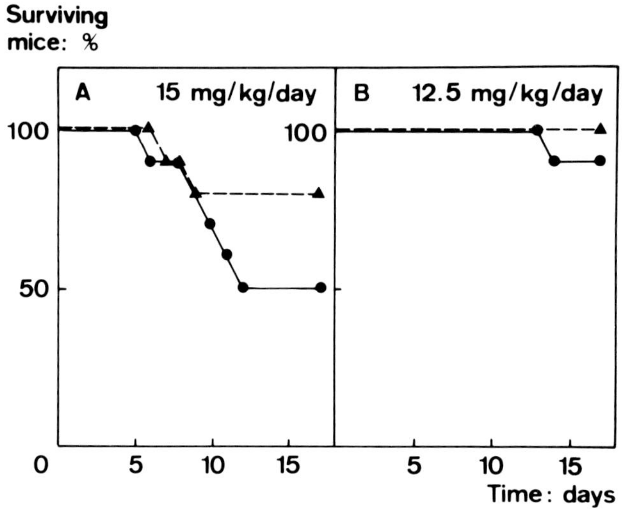

FIGURE 55. Percent of surviving mice after intravenous administration on 2 consecutive days of various doses of free DOX (•); or DOX-adsorbed on PIN (▲). (From Couvreur, P. et al., *J. Pharm. Sci.*, 71, 790, 1982. With permission.)

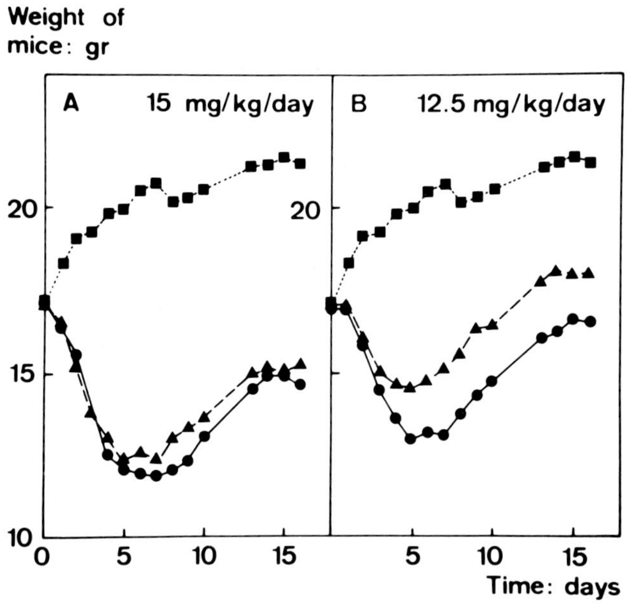

FIGURE 56. Change in body weight with respect to time in mice injected intravenously on 2 consecutive days with various doses of free DOX (•); or DOX adsorbed on PIN (▲). (■) Represents the weight of control mice. (From Couvreur, P. et al., *J. Pharm. Sci.*, 71, 790, 1982. With permission.)

Four days after the last injection (Figure 58), the decrease in total body weight of the doxorubicin-treated mice (as compared to the weight of the control mice) was less in the PIN-DOX group (12%) than in the DOX group (23%), confirming the above-presented data. The same pattern held true for each organ weighed, with the difference in organ weight loss (between DOX and PIN-DOX) being especially marked in the

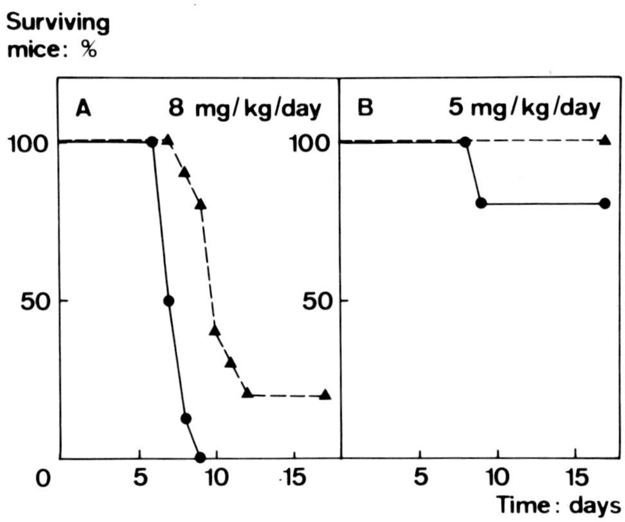

FIGURE 57. Percent of surviving mice after intravenous administration on 5 consecutive days of various doses of free DOX (•); or DOX-adsorbed on PIN (▲). (From Couvreur, P. et al., *J. Pharm. Sci.*, 71, 790, 1982. With permission.)

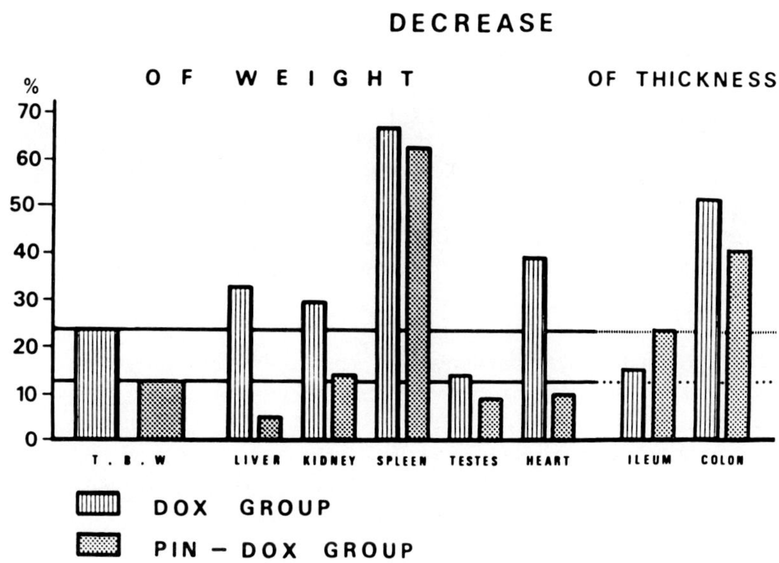

FIGURE 58. Decrease in total body weight, weight of certain organs, and thickness of intestinal mucosa in mice injected intravenously with free DOX and DOX-adsorbed on PIN, at a dose of 7.5 mg of drug/kg body weight/day × 3 consecutive days. Each vertical bar represents the difference between control mice and treated mice values in % of control mice value, the 4th day after the last injection.

heart (29%), liver (28%), and kidneys (16%). It should be emphasized that, in the DOX group, the weight loss in the heart and liver was proportionally higher than the total body weight loss, whereas the opposite was true for the PIN-DOX group. In this respect, it could be noted that, as described in Section IV., the uptake of doxorubicin by the heart is significantly less when the drug is in its adsorbed form than when it is in its free form.

Regarding intestinal mucosa thickness (Figure 58), there was less of a decrease (as compared to the thickness of mucosa in the control group) in the PIN-DOX group (41%) than in the DOX group (52%) in the colon, whereas the opposite was true for the ileum (15 and 22% for DOX and PIN-DOX, respectively); further experiments will be needed to explain these observations.

Taking the megakaryocyte count of the control group as 100%, the counts of the DOX and PIN-DOX groups were 21 and 58% respectively, the 4th day after the last injection. Thirteen days after the last injection, the megakaryocyte count of the DOX group was higher than that of the control group (117%), whereas the count of the PIN-DOX group remained unchanged (58%).

The other histopathologic findings supported the more quantitative, above described observations. In general, symptoms of organ, tissue and cell atrophy were prevalent in the DOX group, with accompanying disturbance of structure, distubance of renewal or growth of cells, signs of altered regeneration, and a few symptoms of degeneration. These alterations were especially marked in the pancreas, liver, kidneys, testes, thymus and bone marrow (Figures 59 and 60).

On the other hand, a disadvantage of PIN-DOX treatment can be seen in the alterations (most likely caused by shortening the infusion rate) it produced in the lungs (Figure 60). Four days after the last injection, these alterations included small foci of emphysema, edema, atelectasis, hyperemia, and swelling of some alveolar septa. In addition, some kind of granular or filamentous foreign bodies (which were surrounded by what appeared to be an endothelial reaction with mononuclear cells) were observed in the lumen of some small blood vessels; 13 days after the last injection, these alterations had become more fibrotic. Further studies will be needed to explain this phenomenon.

VI. APPLICATION OF POLYALKYLCYANOACRYLATE NANOPARTICLES IN EXPERIMENTAL CANCER CHEMOTHERAPY

The purpose of this section is to show whether the use of drug-loaded PAN, as opposed to the use of free drugs, could improve therapeutic efficacy in the treatment of experimental tumors.

A. Dactinomycin-Loaded Nanoparticles

As a preliminary investigation of this question, dactinomycin-loaded polymethylcyanoacrylate nanoparticles (PMN-DACT), one of the first drug-PAN complexes to be worked out, were compared to free dactinomycin (DACT) for activity against soft tissue sarcoma S250, a solid tumor of the LOU/Dec strain of rats whose origin and methods of transplantation have been described elsewhere.[53] This tumor is rather sensitive to dactinomycin treatment.

PMN were prepared by adding the methylic monomer (100 $\mu\ell$) in an aqueous solution of 0.01 *M* HCl (25 mℓ) containing 0.4% polysorbate 20. After polymerization, the suspension was filtered through a fritted glass (pore size 9 to 15 μm) and DACT (2 mg) was then added under stirring. After 30 min, the preparation was buffered to pH 7 using 1.0 *M* NaOH (0.25 mℓ) and a phosphate buffer (4.75 mℓ). The final drug content of the suspension was equal to 66 μg/mℓ while the polymer content was equal to 3.3 mg/mℓ. Sixty-six percent of the added drug was actually bound to the particles under these experimental conditions (see Section II); that corresponded to a drug-to-polymer ratio of about 13 mg/g. The DACT solution used in these assays contained all the above-mentioned reagents, except methylcyanoacrylate monomer. PMN suspensions containing no DACT were also prepared and tested in tumor-bearing animals.

The assays were performed on young adult male rats inoculated subcutaneously in

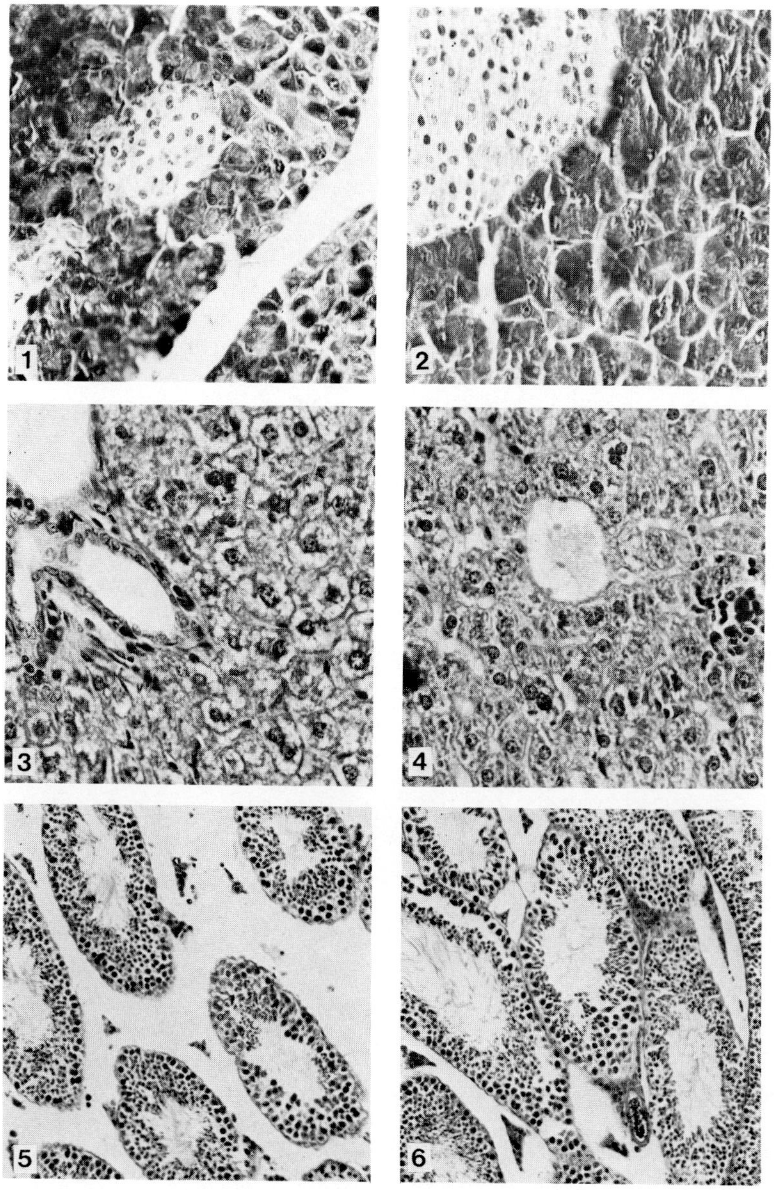

FIGURE 59. Morphologic changes induced in mice treated intravenously with free DOX or with DOX-adsorbed on PIN using an administration schedule of 7.5 mg of drug/kg body weight/day × 3 consecutive days. Scale: the height of each photo number represents 18 μm. (1) Pancreas, DOX group (the 4th day after the last injection). Distinct signs of cellular atrophy: decrease in volume, pyknotic appearance of many nuclei, important reduction or complete disappearance of the cytoplasmic secretion granules in numerous cells. (2) Pancreas, PIN-DOX group (the 4th day after the last injection). The alterations are considerably less pronounced than in the DOX group. (3) Liver, DOX group (the 13th day after the last injection). Diffuse hydropic degeneration of the hepatocytes with distinct "ballooning-cells"-type alteration (mainly in the periportal zone). (4) Liver, PIN-DOX group (the 13th day after the last injection). Small foci of mononuclear infiltration with sporadic eosinocytes, extremely rare acidophilic bodies, and hypertrophic Kupffer cells (the same kind of lesions were less distinct or even nonexistent in the PIN-DOX group). (5) Testis, DOX group (the 4th day after the last injection). Diminished cellularity, degenerated spermatogonia, atrophic seminiferous tubules, and structural decomposition. (6) Testis, PIN-DOX group (the 4th day after the last injection). The changes are considerably less marked than in the DOX group.

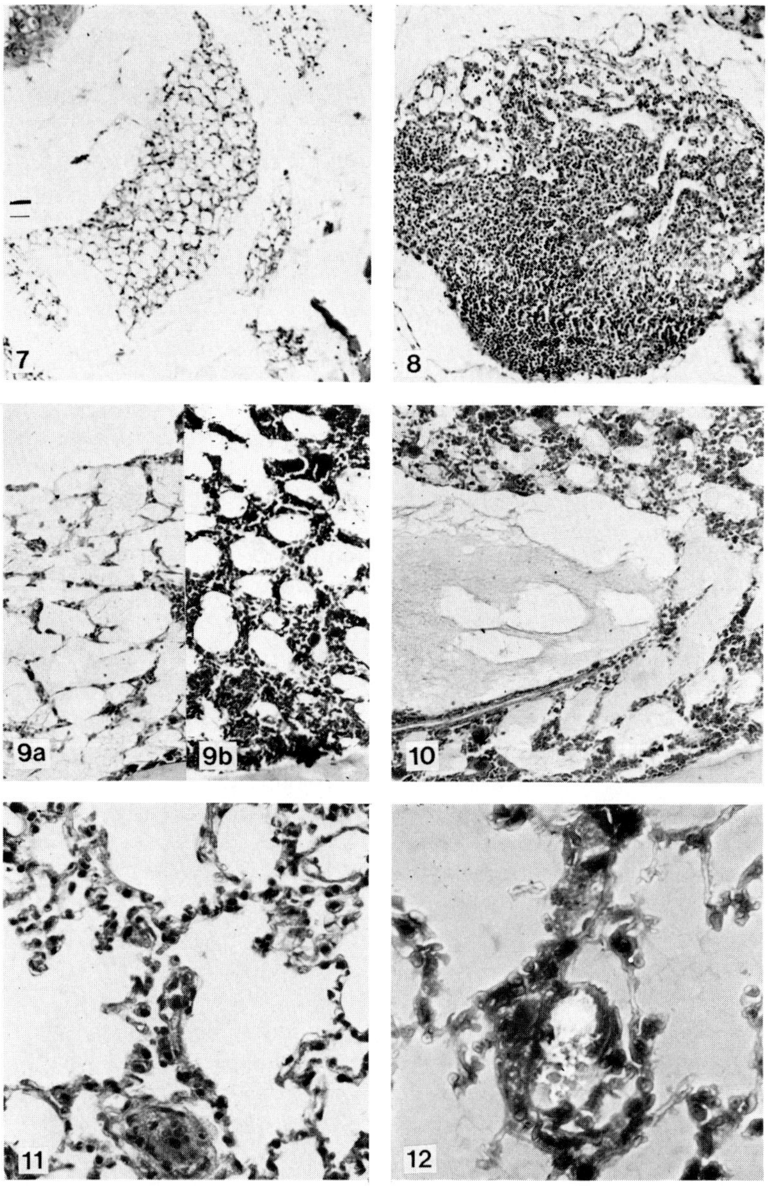

FIGURE 60. A continuation of Figure 59. (7) Thymus, DOX group (the 4th day after the last injection). Nearly complete depletion of the lymphoid tissue (which is replaced by fat). (8) Thymus, PIN-DOX group (the 4th day after the last injection). Less pronounced depletion of atrophic thymocytes than in the DOX group. (9a and 9b) Bone marrow, DOX group (the 4th day after the last injection). Depletion of cellularity and presence of "empty spaces", with noticeable variations among specimens. (10) Bone marrow, DOX group (the 4th day after the last injection). The same alterations as in the DOX group were observed, but with a more regular distribution (on an average, there was no difference in the severity of bone marrow lesions between DOX and PIN-DOX. (11) Lung, DOX group (the 4th day after the last injection). (12) Lung, PIN-DOX group (the 13th day after the last injection). Small foci of foreign-body reaction.

the right flank with 10^6 viable cancer cells harvested from a 2-week-old ascitic form of S250 sarcoma. Ten to twelve days later (i.e., by the time all rats had developed palpable tumor at the site of inoculation), either free or bound DACT was administered by the schedules specified in Tables 14 and 15. The test products were injected in the femoral vein in the ether-anaesthetized animals. Each test group was composed of six rats while another six rats served as a noninjected control in each assay.

Figure 61 shows the evolution pattern of tumor size (estimated with calipers) in two different experiments (exp. No. 1 and 2a). As compared to the tumor increase in the control rats, the tumor growth inhibition in both the DACT and PMN-DACT-treated rats was very apparent. Actually, this effect was much more marked in the case of the PMN-DACT treatment, especially after the second injection. Tests performed with somewhat different injection schedules gave further evidence of the higher activity of the PMN-DACT preparation against sarcoma S250 (Table 14 exp. no. 2b, 2c, and 3). In one case (exp. no. 3), the tumor size ratio between the PMN-DACT and DACT groups was even reduced to zero. Not surprisingly, dactinomycin-free nanoparticles produced no effect on tumor growth (Figure 62 and Table 15).

Based on rat mortality, the bound drug form (PMN-DACT) was found to be more toxic than the free drug (DACT) in these assays (Table 14; exp. nos. 1, 2, 3). An attempt at reducing this undesirable effect was made by using a lower dose. As shown in Table 14 (exp. no. 4a), no deaths occurred among the PMN-DACT-treated rats when the dose administered was reduced by 50% (i.e., 1.65 mℓ instead of 3.30 mℓ of PMN-DACT suspension per kilogram per injection). In addition, unlike the half dose of DACT, the half dose of PMN-DACT still produced an appreciable inhibitory effect on the growth of S250 sarcoma (Figure 63).

Thus, these assays showed that, in comparison to free DACT, PMN-DACT possess a greater efficacy against S250 sarcoma. This enhanced antitumor activity might be explained by better DACT availability for the cancer cells, which might arise from the altered tissue distribution pattern of the drug due to its binding to nanoparticles. If this is the case, then there is also likely to be a higher drug accumulation in healthy tissues susceptible to the adverse effects of DACT, which could account for the higher mortality observed after treatment with the bound drug. The fact that no mortality was noted after the administration of DACT-free nanoparticles (Table 15) partially supports this hypothesis. Moreover, as described in Section IV, the adsorption of DACT to polymethylcyanoacrylate nanoparticles can modify the disposition of the drug in rat tissues to a certain extent.

However, more recently, we succeeded in reducing the higher toxicity of the bound drug form by using an admittedly less histotoxic homologue in preparing the carrier. Indeed, when DACT adsorbed to PHN was tested, it proved to be more effective than the free drug against S250 sarcoma (Figure 64); furthermore, the number of rats surviving beyond 40 days after tumor inoculation was higher after the PHN-DACT treatment (seven survivors out of eight rats) than after the DACT treatment (four survivors out of eight rats). Thus, though more work is necessary before conclusions could be drawn in this area, the increase in toxicity encountered with the PMN-DACT preparation might be overcome by using heavier alkylcyanoacrylate compounds in preparing the dactinomycin-nanoparticles complex. In this respect, recent results obtained by Grislain et al.[54] indicate that DACT-loaded PIN was significantly less toxic to normal mice than the free drug, as expressed by the 1.48-fold higher LD_{50} value observed for the bound drug (upon single i.v. injection).

B. Doxorubicin-Loaded Nanoparticles

1. Activity Against Rat Myeloma S130

The usefulness of DOX-loaded nanoparticles in experimental cancer chemotherapy was investigated at first in LOU/Dec rats bearing S130 myeloma, a tumor line previously used in the study of the antitumor activity of the DOX-DNA complex.[55]

For this purpose, 10^6 viable cancer cells harvested from a LOU rat-bearing ascitic

Table 15
ACTIVITY OF DRUG-FREE PMN IN S250 SARCOMA-BEARING RATS

	Experiment schedule			Tumor size (mm²)[c]				
Exp. no.	Dose[a]	Days of injection[b]	Day of tumor measure[b]	C[d]	PMN	Tumor size ratio PMN/C	t[e]	Mortality[f] PMN
4b	1.65	17 & 18	20	722	611	0.85	0.737	0
			24	1239	1085	0.88	0.701	0
5	3.30	14 & 15	17	1178	1154	0.98	0.185	0
		14 & 16	17	1178	1101	0.94	0.715	0

[a] Volume (mℓ) of PMN per kg body weight per injection.
[b] See Table 14.
[c] Means from six rats.
[d] Control group.
[e] Results of student's *t*-test, indicating no significant difference in tumor size between C and PMN.
[f] See Table 14.

From Brasseur, F., et al., *Eur. J. Cancer,* 16, 1441, 1980. With permission.

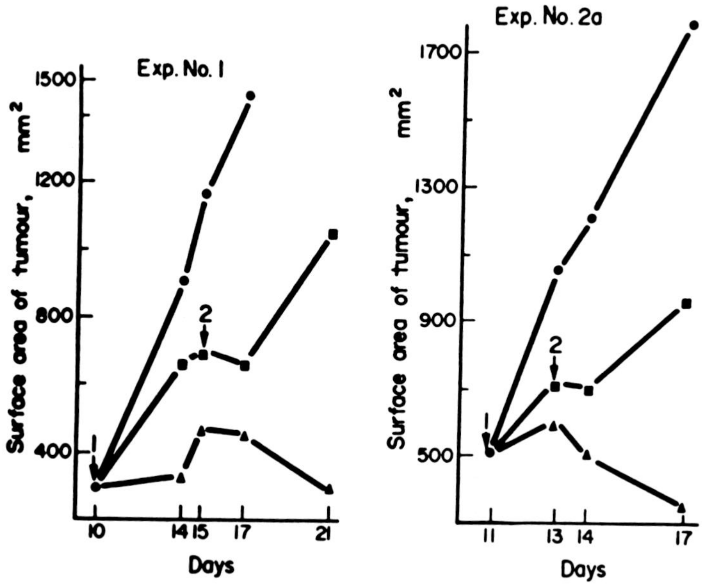

FIGURE 61. Tumor size of S250 treated with free DACT (■); or with PMN-DACT (▲). Dosage: two intravenous injections (↓) of 222 μg of drug (free or adsorbed) per kg body weight (•) represents the tumor size of control rats. (From Brasseur, F. et al., *Eur. J. Cancer,* 16, 1441, 1980. With permission.)

myeloma were transplanted subcutaneously in the flank of each experimental rat. Eight days later (in the first out of two assays) or 12 days later (in the second assay), the animals were given either PIN-DOX or DOX in a single injection in the femoral vein under ether anaesthesia. The polymer content of the PIN-DOX suspension was equal to 10 mg/mℓ. Both the PIN-DOX and DOX preparations contained 1.0 mg of DOX per milliliter and were administered to the animals at a dose of 5 mℓ/kg of body weight. Noninjected tumor-bearing rats served as a control group in each assay. The animals were 5 to 6 in number in each test and control group.

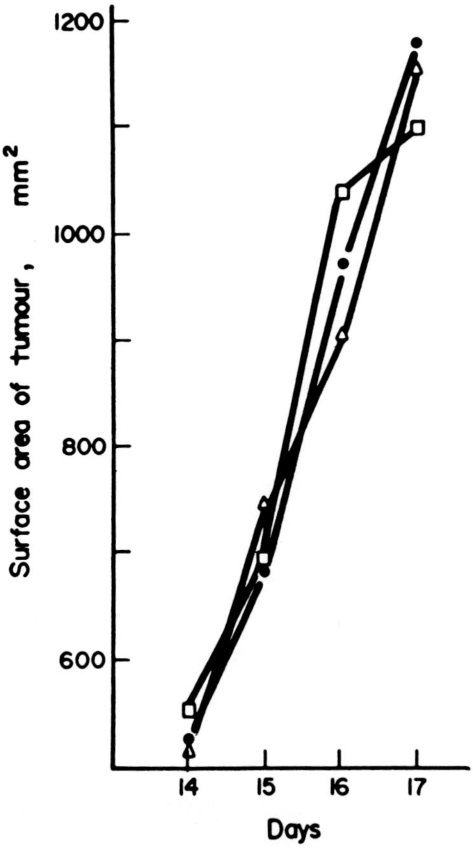

FIGURE 62. Tumor size of S250 after two intravenous injections of drug-free PMN. (△) Injections at days 14, 15; (□) injections at days 14, 16; (•) represents the tumor size of noninjected rats. (From Brasseur, F. et al., *Eur. J. Cancer,* 16, 1441, 1980. With permission.)

As can be seen in Figure 65, the bound drug form exerted an effect on tumor size evolution similar to that of the free drug. In both cases, the tumors started to decrease rapidly from the second day after the injection and had completely disappeared by the 27th to 35th day. By comparison, all control rats with untreated tumors died within the 2 to 3 weeks after tumor transplantation. In fact, S130 myeloma is one of the few well-established transplantable tumors that can undergo complete regression by chemotherapy.[56] However, all rats treated with the free drug died within an 83-day period from tumor inoculation whereas 50% of those rats treated with the bound drug were still alive at the 105th day (Figure 66). Since neither tumor recurrence nor secondary malignancies were observed in any rat, mortality was most likely to be due to the inherent toxicity of DOX. Then we thought that the binding of DOX to PIN could probably reduce the toxicity of the drug without reducing its antitumor activity, and we made further comparisons of the two drug forms with toxicologic and pharmacokinetic studies of healthy animals, as described in preceding sections.

2. Activity Against Murine Leukemia L1210

More recently, we began using the lymphoid L1210 leukemia as a tumor model. In a first assay, BDF1 hybrid male mice were inoculated intravenously with 10^5 leukemic

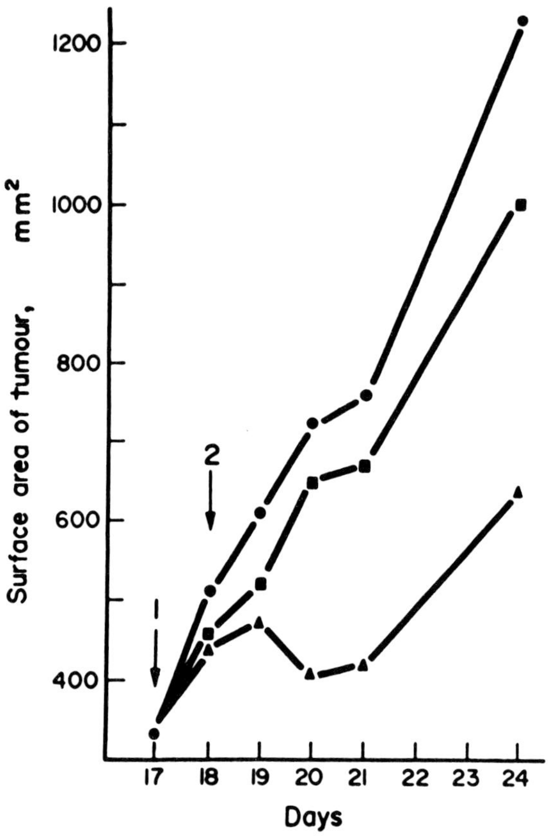

FIGURE 63. Tumor size of S250 treated with free DACT (■); or with PMN-DACT (▲). Dosage: two intravenous injections (↓) of 111 μg of drug (free or adsorbed) per kg body weight (•) represents the tumor size of control rats. (From Brasseur, F. et al., *Eur. J. Cancer,* 16, 1441, 1980. With permission.)

cells and treated during the next 3 days with DOX in either free or bound form. The DOX solution and the PIN-DOX suspension were prepared in the same way as briefly described in the above paragraph. Intravenous doses of 3.0, 6.0, 7.5, and 10.5 mg of drug per kilogram of body weight per day were assayed. Leukemic mice injected with physiological saline were used as a control group. Two dose-zero test groups were also included in the assay, one receiving the free drug vehicle only (i.e., the polymerization medium of the particles), the other the bound drug vehicle only (i.e., drug-free suspension of nanoparticles), following the same schedule of administration as specified above.

The results are presented in Table 16. Based on the survival time of the mice, the adsorbed doxorubicin (PIN-DOX) showed a higher overall effectiveness than the free doxorubicin (DOX) against L1210 leukemia. This enhanced effectiveness became more apparent as the drug dosage was increased, with a significant difference appearing at 3 × 10.5 mg/kg. It is also to be noted that all mice injected with the free drug died within the next 14 days after leukemia inoculation, whereas a 60-day survivor out of nine mice was noted after treatment by 3 × 7.5 mg/kg of the adsorbed drug. As expected, neither the polymerization medium of the particles nor the doxorubicin-free nanoparticles displayed antitumor activity.

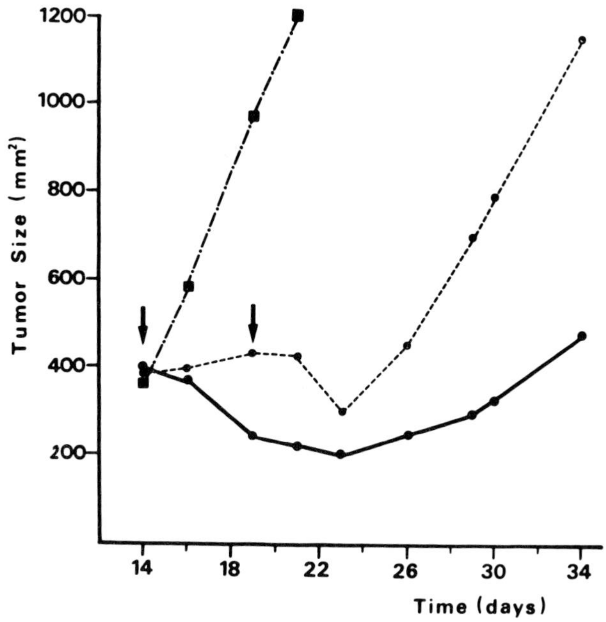

FIGURE 64. Tumor size of S250 treated with free DACT (--O--); or with PHN-DACT (--●--). Dosage: two intravenous injections (↘) of 222 μg of drug (free or adsorbed) per kg body weight (■) represents the tumor size of control rats.

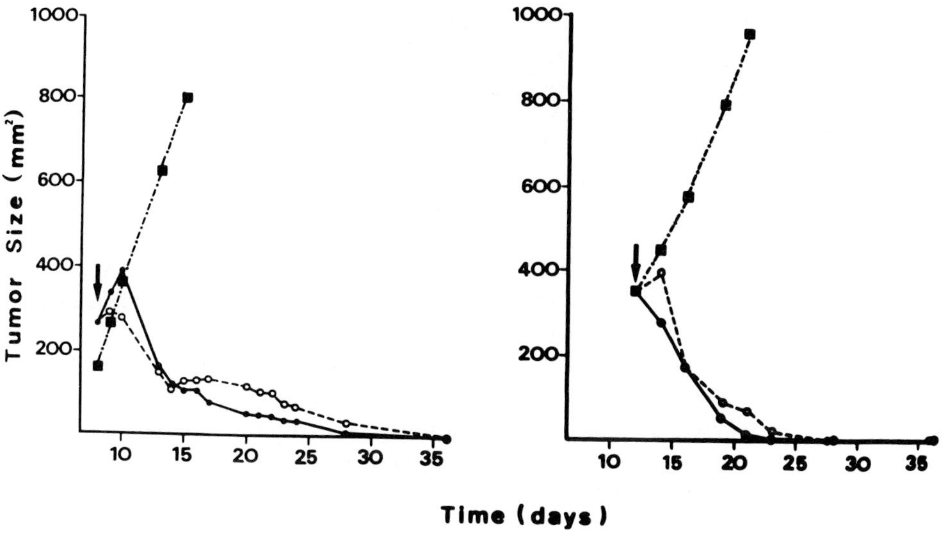

FIGURE 65. Tumor size of S130 treated with free DOX (O); or with PIN-DOX (●). Dosage: one intravenous injection (↘) of 5 mg of drug (free or adsorbed) per kg body weight (■) represents the tumor size of control rats. The results of two separate experiments are presented.

After the administration of the bound drug, the body weight loss of the mice as ascertained on the 5th day was slightly lower, generally, than with the free drug; this may be attributed to the lower chronic toxicity of the bound drug form. This observation is in agreement with the toxicity data obtained in healthy animals (see Section V.). However, as regards acute toxicity, one premature death (5th day) was noted among

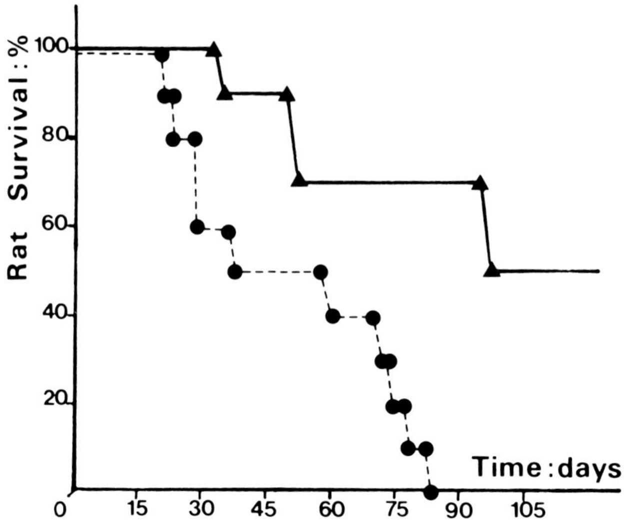

FIGURE 66. Survival rate of S130 myeloma-bearing rats after administration of free DOX (•); or PIN-DOX (▲). Dosage: see caption to Figure 65. Each curve is a composite of the results observed in the two experiments depicted in Figure 65.

the mice treated with the highest dosage of bound drug, whereas the free drug induced no early deaths. Taken as a whole, these preliminary results were considered very encouraging, and we are in the process of confirming them by experimenting with other treatment regimens and drug doses.

VII. CONCLUSIONS

This chapter describes the design of a biodegradable ultrafine drug carrier. Nanoparticles were obtained by emulsion polymerization of alkylcyanoacrylate following an anionic initiation mechanism. The little spheres obtained show a spherical shape and an average diameter of 0.2 μm. However, by simply modifying the acidity of the polymerization medium, it is possible to obtain larger (0.5 μm) or smaller (0.08 μm) particles. These latter would be able to filter through the endothelial wall of capillaries and present a better access to tumors.

Comparative to previously described particulate carriers, polyaklylcyanoacrylate nanoparticles are an original colloidal drug delivery system because they are built by the tangling up of numerous small oligomers rather than by the rolling up of a few macromolecules. Such oligomeric systems could be cleared more easily from the body and be less subject to polymeric overload in the reticuloendothelial system.

A freeze-fracture study of these submicroscopic particles has shown a solid and porous internal structure. Owing to their small size and to their porous nature, nanoparticles are able to adsorb efficiently different kinds of drugs. The most interesting characteristic of polyalkylcyanoacrylate nanoparticles is that they are more or less quickly degraded depending on the length of their alkyl chain. Indeed, the degradable character of polyalkylcyanoacrylate has been demonstrated in vitro and in vivo by many authors. These investigations generally reveal a degradation mechanism based on the hydrolytic rupture of the hydrocarbon chain of the polymer. This degradation scheme which corresponds to a reverse Knoevenagel reaction, yields formaldehyde and cyanoacetate. By a quantitative estimation of the formaldehyde production, we suc-

Table 16
COMPARISON OF FREE DOX WITH PIN-DOX IN CHEMOTHERAPY OF L-1210 LEUKEMIA[a]

	Volume administered per mouse of 25 g (mℓ/injection)	Dosage (mg/kg/injection)		Body weight change (g on day 5)	Number of day-5 survivors	Mean survival time (days)	T/C[c]	Numbers of survivors	
		Drug	Carrier					Day 30	Day 60
Control (saline)[b]	0.35			+1.0	30/30	7.6±0.49[d]	—	0/30	
DOX[b]	0.35	0.0		+0.3	9/9	7.4	100	0/9	
	0.10	3.0		−1.0	9/9	9.4	129	0/9	
	0.20	6.0		−2.8	9/9	10.5	143	0/9	
	0.25	7.5		−4.3	9/9	12.1	171	0/9	
	0.35	10.5		−4.9	9/9	10.4	143	0/9	
PIN-DOX[b]	0.35	0.0	140	+0.1	9/9	8.0	114	0/9	
	0.10	3.0	40	−0.8	9/9	9.4	129	0/9	
	0.20	6.0	80	−2.6	9/9	12.1	171	0/9	
	0.25	7.5	100	−3.2	9/9	14.5[e]	200[e]	1/9	1/9
	0.35	10.5	140	−4.2	8/9	15.0	229	0/9	

[a] BDF1 male mice were inoculated iv on day 0 with 10^5 leukemic cells from a DBA2 mouse bearing 7-day-old ascitic L1210.
[b] One iv injection per day during three consecutive days (1,2,3).
[c] T/C = median survival time of treated mice/median survival time of control mice, in percentage.
[d] Mean ± SD.
[e] Exclusive of the one long-term survivor.

ceeded in showing that this degradation mechanism plays only a secondary role. On the other hand, the large production of isobutanol and acid observed during the degradation of PIN demonstrates the likeliness of a degradation mechanism by hydrolysis of the polymers ester function. According to this reaction mechanism, the hydrophilicity of the hydrocarbon chain rises progressively until the polymer is solubilized in various aqueous media. Besides, various tests conducted in the presence of rat hepatic microsomes or tritosomes pointed out the ability of this degradation mechanism to undergo an enzymatic catalysis.

Intravenously administered PAN are rapidly cleared from the blood. The carrier concentrates in organs rich in reticuloendothelial cells (namely the liver, spleen, and bone marrow). After longer time intervals after the injection, the liver still retains the major part of the nanoparticles as compared to other organs. However, total liver radioactivity gradually decreases and the excretion of nanoparticles is complete after 7 days, mainly via the urine (60%) and the feces (40%). The situation can thus be summarized by the assertion that the liver acts as a reservoir towards nanoparticles, conditioning their quick first phase disappearance from the blood and their second phase excretion from the body under degraded form.

Owing to the high proportion of the carrier taken up by the liver, it is also probable that this organ represents the major metabolization site for these polymeric particles. The Kupffer cells are the major site of internalization of the nanoparticles irrespective to their size and following a very quick capture mechanism. Endothelial cells show a slower and lower uptake, and parenchymal cells a very low uptake.

The most acute problem of colloidal carrier such as nanoparticles still remains the passage through the endothelial barrier to other sites than the liver, spleen, or bone marrow. However, we demonstrated the possibility of such a passage of polyisobutylcyanoacrylate nanoparticles to a neoplastic tissue. Indeed, whole body autoradiography performed on Lewis lung carcinoma-bearing mice showed progressive accumulation of the carrier in the primary tumor; furthermore, a high level of radioactivity was found in the metastasized lungs of cancerous animals whereas no radioactivity accumulation was observed in the lungs of healthy mice. This phenomenon could be a valuable reason to justify the use of nanoparticles as a drug carrier in the treatment of pulmonary metastases. It has been shown that the binding of cytostatic drugs to nanoparticles modifies their distribution patterns in animals and generally increases the tissue capture of these drugs. Furthermore, the use of polyalkylcyanoacrylate nanoparticles as a drug carrier can considerably reduce the inherent toxicity and side effects of cytotoxic drugs.

Preliminary results indicate that the use of polymethylcyanoacrylate nanoparticles as a drug carrier enhances the cytostatic activity of the drug towards a transplanted soft tissue sarcoma of the rat. More recently, we began using the lymphoid L1210 leukemia of the mouse as a tumor model and we obtained very encouraging results with doxorubicin-loaded polyisobutylcyanoacrylate nanoparticles.

Finally, the first toxicological data obtained at the cellular and whole-body levels did not show any acute or subacute toxicity likely to restrict the use of polyalkylcyanoacrylate nanoparticles in human medicine.

From our point of view the future of polyalkylcyanoacrylate nanoparticles cannot be dissociated from that of other colloidal carriers. The main disadvantage of all colloidal carriers is their marked tendency to concentrate in the liver, especially in Kupffer cells, leading to a decrease in the availability of the adsorbed drugs towards other tissues. In some cases, this character can be of interest, for example in the treatment of infectious diseases of the liver (leishmaniasis, cryptococcosis, histoplasmosis). However, modifications in the tissue distribution of an adsorbed drug and its release from a "storage" organ can decrease the drug side-effects, especially when this organ consists of tissues which are not sensitive to the toxic effects of the drug. If the cytostatic agent acts at specific phases of the cell cycle, an increase in the time that the cancerous tissue is exposed to the drug will be beneficial.

ACKNOWLEDGMENTS

This work was supported by the SOPAR SA Company and by the IRSIA (Institut pour l'Encouragment de la Recherche Scientifique dans l'Industrie et l'Agriculture). The authors wish also to thank the FNRS (Fonds National de la Recherche Scientifique) and the AGF (Assurances Génèrales de France) for their kind support.

REFERENCES

1. Rhaman, Y. E., Cerny, E. A., Tollasken, S. L., Wright, B. J., Nance, S. L., and Tompson, J. F., Liposome-encapsulated Actinomycin D: potential in cancer chemotherapy, *Proc. Soc. Exp. Biol. Med.,* 146, 1173, 1974.
2. Gregoriadis, G., Homing of liposomes to target cells, *Biochem. Soc. Trans.,* 3, 613, 1975.
3. Birrenbach, G. and Speiser, P., Polymerised micelles and their use as adjuvants in immunology, *J. Pharm. Sci.,* 65, 1763, 1976.
4. Couvreur, P., Tulkens, P., Roland, M., Trouet, A., and Speiser, P., Nanocapsules: a new type of lysomotropic carrier, *Feb. Lett.,* 84, 323, 1977.
5. Couvreur, P., Kante, B., Roland, M., Guiot, P., Baudhuin, P., and Speiser, P., Polycyanoacrylate nanocapsules as potential lysosomotropic carriers: preparation, morphological and sorptive properties, *J. Pharm. Pharmacol.,* 31, 331, 1979.
6. Wilkinson, T. S., Tissue adhesives in cutaneous surgery, *Arch. Dem.,* 106, 834, 1972.
7. Donnelly, E. F., Johston, D. S., Peffer, D. C., and Dunn, D. J., Ionic and zwitterionic polymerization of n-alkyl 2-cyanoacrylates, *Polym. Lett. Ed.,* 15, 399, 1977.
8. Couvreur, P., Kante, B., Roland, M., and Speiser, P., Adsorption of antineoplasic drugs to polyalkylcyanoacrylate nanoparticles and their release characteristics in a calf serum medium, *J. Pharm. Sci.,* 68, 1521, 1979.
9. Moor, H., Multethaler, K., Waldner, H., and Frey-Wissling, A., A new freezing-ultramicrotome, *J. Biophys. Biochem. Cytol.,* 10, 1, 1961.
10. Wicksell, S. D., The corpuscle problem. A mathematical study of a biometric problem, *Biometrika,* 18, 151, 1925.
11. Guiot, P., Aspects Quantitatifs de l'Interaction entre Vecteurs Particulaires et Cellules, Thèse de Doctorat, Université Catholique de Louvain, Louvain, Belgium, 1981.
12. Guiot, P., Baudhuin, P., and Gotfredsen, C. F., Morphological characterisation of liposome suspensions by stereological analysis of freeze-fracture replica from spray-frozen samples, *J. Microsc.,* 120, 159, 1980.
13. Guiot, P. and Baudhuin, P., Size distribution analysis of liposomes by electron microscopy, in *Liposome Technology,* Gregoriadis, G., Ed., CRC Press, Boca Raton, Fla., 1984, 163.
14. Kante, B., Couvreur, P., Lenaerts, V., Guiot, P., Baudhuin, P., and Speiser, P., Tissue distribution of 3H-Actinomycin-D adsorbed on polybutylcyanoacrylate nanoparticles, *Int. J. Pharm.,* 7, 45, 1980.
15. Guiot, P. and Couvreur, P., Quantitative study of the interaction between polybutylcyanoacrylate nanoparticles and mouse peritoneal macrophages in culture, *J. Pharm. Belge,* 38, 130, 1983.
16. Leonard, F., Kulkarni, R., Brandes, G., Nelson, J., and Cameron, J., Synthesis and degradation of poly(alkylcyanoacrylates), *J. Appl. Polym. Sci.,* 10, 259, 1983.
17. Vezin, W. and Florence, A., In vitro degradation rates of biodegradable poly-N-alkylcyanoacrylates, *J. Pharm. Pharmacol.,* 30, 5, 1978.
18. Grislain, L., Couvreur, P., Lenaerts, V., Roland, M., Deprez-De Campeneere, D., and Speiser, P., Pharmacokinetics and distribution of a biodegradable drug carrier, *Int. J. Pharm.,* 15, 335, 1983.
19. Wade, D. and Leonard, F., Degradation of poly(methyl-2-cyanoacrylates), *J. Biomed. Mat. Res.,* 6, 215, 1972.
20. Trouet, A., Isolation of modified liver lysosomes, *Meth. Enzymol.,* 31, 323, 1974.
21. Lenaerts, V., Couvreur, P., Christiaens-Leyh, D., Joiris, E., Roland, M., Rollman, B., and Speiser, P., Degradation of poly(isobutylcyanoacrylate) nanoparticles, *Biomaterials,* 5, 65, 1984.

22. Leyh, D., Couvreur, P., Lenaerts, V., Roland, M., and Speiser, P., Etude du mecanisme de dégradation des nanoparticules de polycyanoacrylate d'alkyle, *Labo-Pharma,* 32, 100, 1984.
23. Lowry, O. H., Rosebrough, N. J., Farr, A. L., and Randall, R. J., Protein measurement with the folin phenol reagent, *J. Biol. Chem.,* 193, 265, 1951.
24. Steinman, R. M., Brodies, S. E., and Cohn, Z. A., Membrane flow during pinocytosis. A stereological analysis, *J. Cell. Biol.,* 68, 665, 1976.
25. Ullberg, S., Studies on the distribution and fate of S 35 labelled benzyl-penicillin in the body, *Acta Radiol.,* 118, 1, 1954.
26. Nagelkerke, J., Barto, K., and Van Berkel, T., in *Sinusoidal Liver Cells,* Knook, E. and Wisse, E., Eds., Elsevier, Amsterdam, 1983, 319.
27. Couvreur, P., Kante, B., Lenaerts, V., Scailteur, V., Roland, M., and Speiser, P., Tissue distribution of antitumor drugs associated with polyalkylcyanoacrylate nanoparticles, *J. Pharm. Sci.,* 69, 199, 1980.
28. Coover, H. N., Joyner, F. B., Shearer, N. H., Wicker, T. H., Chemistry and performance of cyanoacrylate adhesives, *Soc. Plastics Eng. J.,* 15, 413, 1959.
29. Woodward, S. C., Herrmann, J. B., Cameron, J. L., Brandes, G., Pulaski, E. I., and Leonard, F., Histotoxicity of cyanoacrylate tissue adhesives in the rat, *Ann. Surg.,* 162, 113, 1965.
30. Pani, K. C., Gladieux, G., Kulkarni, R. K., and Leonard, F., The degradation of N-butyl alpha cyanoacrylate tissue adhesive, *Surgery,* 63, 481, 1968.
31. Collins, J. A., James, P. M., Levitsky, S. A., Bredenburg, C. E., Anderson, R. W., Leonard, F., and Hardaway, R. M., II, Clinical use in severe combat casualties. Cyanoacrylate adhesive as topical hemostatic aids. II., *Surgery,* 65, 260, 1969.
32. Matsumoto, T. and Hardaway, R., Cyano-acrylate adhesive and hemostasis, *Arch. Surg.,* 94, 858, 1967.
33. Kante, B., Couvreur, P., Dubois-Krack, G., De Meester, C., Guiot, P., Roland, M., Mercier, M., and Speiser, P., Toxicity of polyalkylcyanoacrylate nanoparticles I: Free nanoparticles, *J. Pharm. Sci.,* 7, 786, 1982.
34. Ames, B. N., Mc Cann, J., and Yamasaki, E., Methods for detecting carcinogens and mutagens with the Salmonella/mammalian-microsome mutagenicity test, *Mutat. Res.,* 31, 347, 1975.
35. Ames, B. N., Durston, W. F., Yamasaki, E., and Lee, F. D., Carcinogens are mutagen: a simple test system combining liver homogenates for activation and bacteria for detection, *Proc. Nat. Acad. Sci. U.S.A.,* 70, 2281, 1973.
36. Ames, B. N., Lee, F. D., and Durston, W. E., An improved bacterial test system for the detection and classification of mutagens and carcinogens, *Proc. Natl. Acad. Sci. U.S.A.,* 70, 782, 1973.
37. Ames, B. N., Kammen, H. O., and Yamasaki, E., Hair dyes are mutagenic: identification of a variety of mutagenic ingredients, *Proc. Natl. Acad. Sci. U.S.A.,* 72, 3423, 1975.
38. Collins, J. A., Pani, C., Seidenstein, M., Brandes, G., and Leonard, F., Cyanoacrylate adhesive as topical hemostatic aids. I. Experimental evaluation on liver wounds in dogs, *Surgery,* 65, 256, 1969.
39. Arianoff, A., Vielle, G., and Arianoff, V., Autoplasies des eventrations par greffe de peau totale pédiculée, *Acta Chir. Belg.,* 78, 325, 1979.
40. Couvreur, P., Kante, B., and Roland, M., Les perspectives d'utilisation des formes microdispersées comme vecteurs intracellulaires, *Pharm. Acta Helv.,* 53, 341, 1978.
41. Lenaerts, V., Nagelkerke, J. F., Van Berkel, T. J. C., Couvreur, P., Grislain, L., Roland, M., and Speiser, P., In vivo uptake of poly-isobutylcyanoacrylate nanoparticles by rat liver Kupffer, endothelial and parenchymal cells, *J. Pharm. Sci.,* 73, 980, 1984.
42. Lenaerts, V., Contribution à l'Étude de Dégradation et de la Distribution des Nanoparticules de Polycyanoacrylate d'alkyle, Ph.D. thesis, Louvain-en-Woluwe, 1984.
43. Rondelet, J. A. and Froment, P., Synthèse et propriétés de la N-(hydroxy-2-éthyl)-4 piperazinyl carboxymethyl-1 tetracycline, *Ann. Pharm. Franc.,* 27, 63, 1968.
44. Couvreur, P., Mise au Point d'un Nouveau Vecteur de Médicament, Thèse d'agrégation de l'enseignement supérieur, Louvain-en-Woluwe, 1983.
45. Carter, S. K., Adriamycine — a review, *J. Natl. Cancer Inst.,* 55, 1265, 1975.
46. Blum, R. H. and Carter, S. K., Adriamycine — a new anticancer drug with significant clinical activity, *Ann. Intern. Med.,* 80, 249, 1979.
47. Wang, J. J., Cortes, E., Sinks, L. F., and Holland, J. F., Therapeutic effect and toxicity of adriamycin in patients with neoplastic disease, *Cancer,* 28, 837, 1971.
48. Lefrak, E., Pitha, J., Rosenheim, S., Gottlieb, J., Clinicopathologic analysis of adriamycin cardiotoxicity, *Cancer,* 32, 302, 1973.
49. Trouet, A., Deprez-De Campeneere, D., De Smedt-Malengreaux, M., and Atassi, G., Experimental leukemia chemotherapy with a "lysomotropic" adriamycin-DNA complex, *Eur. J. Cancer,* 10, 405, 1974.

50. Trouet, A. and Deprez De Campeneere, D., Daunorubicin-DNA and Doxorubicin-DNA: a review of experimental and clinical data, *Cancer Chemother. Pharmacol.*, 2, 77, 1979.
51. Olson, F., Mayhew, E., Maslow, D., Rustum, Y., and Szoka, F., Characterization of Adriamycin encapsulated in liposomes, *Eur. J. Cancer Clin. Oncol.*, 18, 167, 1982.
52. Couvreur, P., Kante, B., Grislain, L., Roland, M., and Speiser, P., Toxicity of polyalkylcyanoacrylate nanoparticles. II. Doxorubicin loaded nanoparticles, *J. Pharm. Sci.*, 71, 790, 1982.
53. Brasseur, F., Couvreur, P., Kante, B., Deckers-Passeau, L., Roland, M., Deckers, C., and Speiser, P., Actinomycin-D adsorbed on polymethyl cyanoacrylate nanoparticles: increased efficiency against an experimental tumor, *Eur. J. Cancer*, 16, 1441, 1980.
54. Grislain, L., Couvreur, P., and Van Snick, L., submitted, 1984.
55. Deckers, C., Mace, F., Deckers-Passau, L., and Trouet, A., in *Molecular Base of Malignancy*, Deutsh, E., Moser, K., Rainer, H., and Stacher, A., Eds., Geory Thieme Publ., Stuttgard, 1976, 254.
56. Deckers, C., Deckers-Passau, L., and de Halleux, F., Regression of transplanted immunoglobulin-secreting rat tumors by irradiation and chemotherapy and induction of transplantation resistance, *Cancer Res.*, 33, 2338, 1973.

Chapter 3

BIOMEDICAL APPLICATIONS OF MAGNETIC DRUG CARRIERS IN CANCER CHEMOTHERAPY*

Kenji Sugibayashi and Yasunori Morimoto

TABLE OF CONTENTS

* Color Figures 23 and 24 appear following page 120.

I. INTRODUCTION

Together with surgery, radiotherapy, and immunotherapy, chemotherapy is one of the most frequently used treatment modalities in cancer therapy. The earlier we find the tumor and remove it surgically, the better therapeutic efficacy becomes. However, surgery is severely limited in that it is incapable of curing patients with disease that has spread beyond the area of primary treatment. Chemotherapy in the treatment of such metastatic disease and in the prevention of tumor recurrences becomes important in the majority of cancer patients who cannot be cured by surgery or radiotherapy. In addition, to cure the tumor without changing the shape of organs is a final aim in cancer therapy.

Since most antitumor agents which are available in clinical stage work not only on tumor cells but also on normal cells, they affect the survival of the normal host tissues too. After antitumor agents are administered systemically, they may be distributed generally throughout the body the same as other drugs. Such shortcomings reduce the efficacy of cancer chemotherapy. Localization of chemotherapeutic agents to specific sites would greatly improve the potential of cancer chemotherapy. Much more attention would be paid to the targeting of chemotherapeutic agents into the tumor sites from that point.

Recently, several types of drug carriers intended to modify the systemic distribution of chemotherapeutic agents have been proposed as possible drug delivery systems. The liposome carrier system for cancer therapy was the focus of many authors.[1,2] It was reported that liposomes were taken up selectively into the reticuloendothelial system of the liver and spleen after intravenous (i.v.) injection.[3,4] The mechanism of "endocytosis" has been discussed using liposomes containing antitumor agents.[5,6] Trouet et al.[7,8] reported that deoxyribonucleic acid (DNA) showed a tight affinity with antitumor agents, (i.e., daunomycin, adriamycin) and that DNA-drug complexes reduced toxicity and increased effectiveness in the treatment of mouse leukemia. The complexes were easily taken up into a cell by endocytosis and digested with lysosomal enzymes, with resulting free-active agents released into a cell. We recently reported that 5-fluorouracil trapped in albumin microspheres (nonmagnetic microspheres) was present in high levels in the liver of mice after i.v. injection[9,10] and suggested the sustained release and prolonged action of the trapped drug in mice beared Ehrlich ascites[11] and solid carcinoma.[12] In addition, adriamycin entrapped in albumin microspheres showed pronounced antitumor activity on AH 7974 liver metastasis in rats as a model with an experimental tumor after intraportal injection.[13] However, i.v. injection of those drug carriers as mentioned above results in their uptake predominantly by the reticuloendothelial system.[5,9,10]

A few researchers have reported some biomedical applications of magnet and/or magnetic drug carriers in cancer therapy. For example, externally guided ferromagnetic silicone has been used experimentally to accomplish selective vascular occlusion and necrosis of hypernephromas in man.[14] Kato[15] already suggested that ferromagnetic ethylcellulose microcapsules containing mitomycin C could be guided in VX2 tumors in a rabbit's limb and urinary bladder under the usual magnetic field of a conventional magnet. Widder et al.[16] also suggested that, since magnetic microspheres injected into the rentral caudal artery could be localized to some extent to a predetermined target site by an externally applied magnetic field, magnetically guided albumin microspheres containing a drug would be useful as a drug delivery system with site specificity. In addition, Mosbach and Schroder[17] prepared magnetic starch microspheres containing carbonyl-iron powder and tried targeting one part of a rabbit's ear. More recently Ibrahim et al.[18] prepared magnetically responsive polyalkylcyanoacrylate nanoparticles containing dactinomycin and suggested a possibility of localizing a high concentration of drug in a desired target and another possibility of reducing accumulation in the reticuloendothelial system with the nanoparticles. If site specific drug delivery of antitumor agents could be achieved with magnetic means as shown above, this delivery system would eliminate adverse side effects that are often the sequelae of generated systemic drug distribution. This chapter describes mainly the utility of the magnetic albumin microspheres (ferro-colloid-entrapped albumin microspheres) and the magnetic emulsions (oil-in-water emulsions) which we and our collaborators have investigated.[19-26]

II. MAGNETIC MICROSPHERES

Targeting of albumin microspheres containing antitumor agents and magnetites using magnetic field is achieved as follows. As shown in Figure 1, after magnets are applied to produce a magnetic field on a target site such as a tumor site and magnetically responsive albumin microspheres are administered into the vascular system, the microspheres may concentrate in the target site and release the entrapped drug there, resulting in many antitumor agent molecules distributing near the target organs. Antitumor agents preferentially distributed about the target tumor cells would work much better than systemically distributed drugs.

A. Preparation of Magnetic Albumin Microspheres

Magnetic albumin microspheres can be prepared with serum albumin and magnetic fluid. The magnetic fluids are technical materials which have characteristics like a fluid (liquid) and magnetic properties like iron powders.[27] Fe_3O_4 colloidal particles are evenly dispersed in several solvents, (i.e., water, chloroform, kerosene) with the diameter of Fe_3O_4 particles in magnetic fluids about 100 to 200 Å. Each one layer of sodium oleate and dodecylbenzene sulfate adsorbs on the surface of Fe_3O_4 particles for water-based magnetic fluids. Several kinds of magnetic fluids are on the market today. On the other hand, the smallest Fe_3O_4 powder which we can easily buy is about 1 μm in diameter. Repeated grind may produce smaller powders, but they are not enough to make the magnetic microspheres sized below 2 μm in diameter. They may be used for making large albumin spheres. An example for preparation of water-based magnetic fluid is shown in Chart 1.

The magnetic albumin microspheres are prepared by phase separation emulsion polymerization.[28] A mixed aqueous solution (2.5 mℓ) containing 150 to 200 mg of serum albumin (i.e., bovine or human serum albumin, Fraction V), 100 to 250 mg of antitumor agents such as fluorouracil 5 and adriamycin, and 0.5 mℓ of water-based magnetic fluid was emulsified into 100 mℓ of cottonseed oil containing 10% (v/v) Span 85 with a glass stirrer attached to an electric motor (Tokyo Rikakikai, Model MS-75) at 2500 rpm for 10 min and with a high speed stirrer (Ultra-Turrax, Janke & Kunkel, W. Germany) at 8000 or 20,000 rpm for 30 min at 20°C. Isooctane/chloroform or n-octane may be used instead of cottonseed oil/Span 85.[29] The resulting water (albumin aqueous solution) in oil emulsions can be solidified using some chemicals and/or heat. In these experiments, the microspheres containing 5-fluorouracil and adriamycin were solidified at 180 and 120°C, respectively. After cooling the reaction flask containing magnetic microspheres and cottonseed oil, the microspheres were extracted from the oil using an organic solvent such as ether. The magnetic microspheres were stored in a desiccator at 4°C. Apparatus and flow chart are shown in Figure 2 and Chart 2, respectively.

We prepared 5-fluorouracil-, adriamycin-, mitomycin C-, and sulfanilamide-entrapped magnetic and/or nonmagnetic microspheres. Kramer[30] and Lee et al.[31] have used mercaptopurine and progesterone, respectively. From these experiments, it is suggested generally that heat-resistant and water soluble drugs are suitable for entrapping

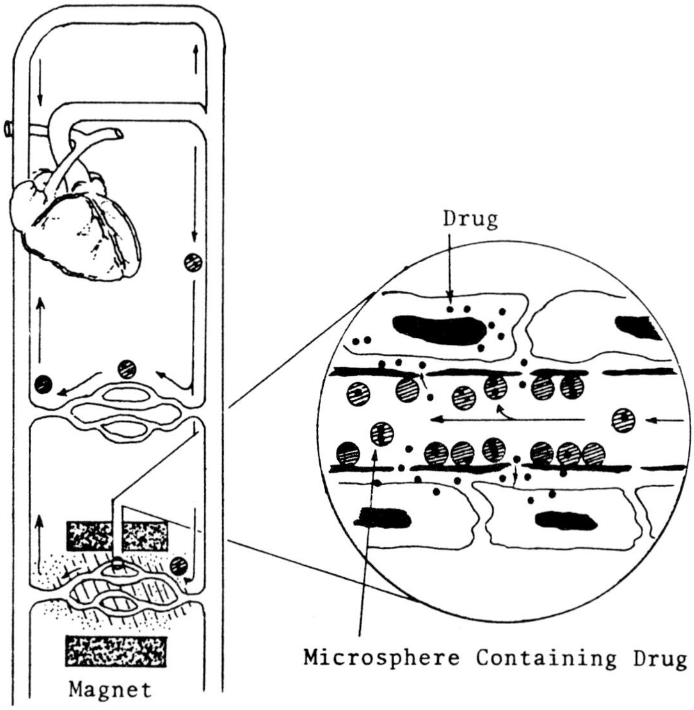

FIGURE 1. Schematic representation of the possible mechanism for drug delivery to target tissues by magnetic albumin microspheres after i.v. administration.

agents in albumin microspheres. For unstable drugs against heat, however, it is possible to entrap albumin microspheres with carbonyl agents[32,33] and electrolytes.[34] For drugs which are practically insoluble in water, some solubilizers may be used and chemical bonds may be applied between drugs and previously prepared nonentrapped albumin microspheres.[35] Combined with several similar methods, not only small molecules but also enzymes and antibodies could be entrapped in albumin microspheres.

B. Physicochemical Properties and Magnetic Characterization of Magnetic Microspheres

1. Physicochemical Properties

The shape of albumin microspheres was invariably spherical probably due to the preparation method where the inner albumin phase of w/o emulsion was immediately solidified at high temperature. As shown in a scanning electron micrograph (Figure 3), however, the microspheres have rough surface. Compared with the diameter of magnetites (100 to 200 Å, see Figure 4), the size of these projections is almost the same. It suggested that magnetite powders may arrange on the surface of albumin microspheres. The size of magnetic microspheres may be changed with a composition in the inner phase of w/o emulsions at the first step of preparation. Strength of stirring and solidification temperature also are important parameters for the diameter of microspheres. We have made several kinds of microspheres which have average diameters of 0.2 to 50 μm. In these experiments we mainly use 1 μm- and 3 μm-microspheres.

2. Toxicity of Magnetic Microspheres

Toxicity of magnetic albumin microspheres can be discussed in two parts; toxicity of (nonmagnetic) albumin microsphere itself and toxicity due to ferrite. It is very dif-

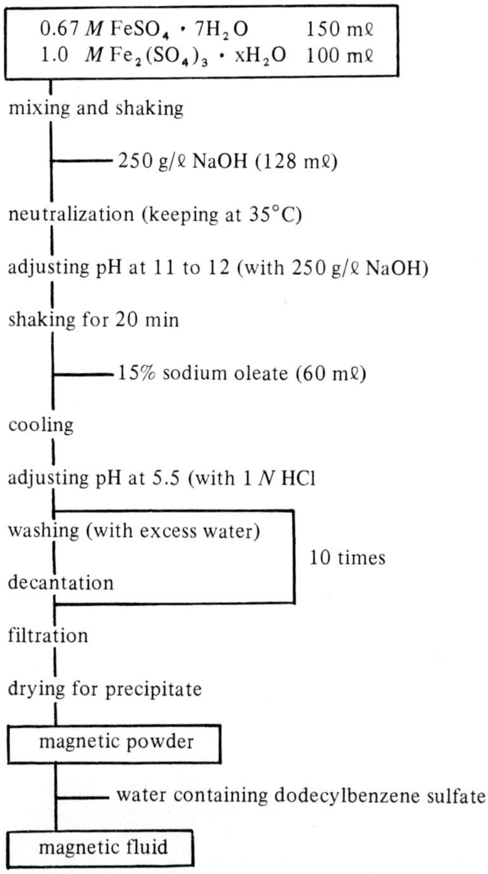

CHART 1. Schematic diagram of preparation of magnetic fluids.

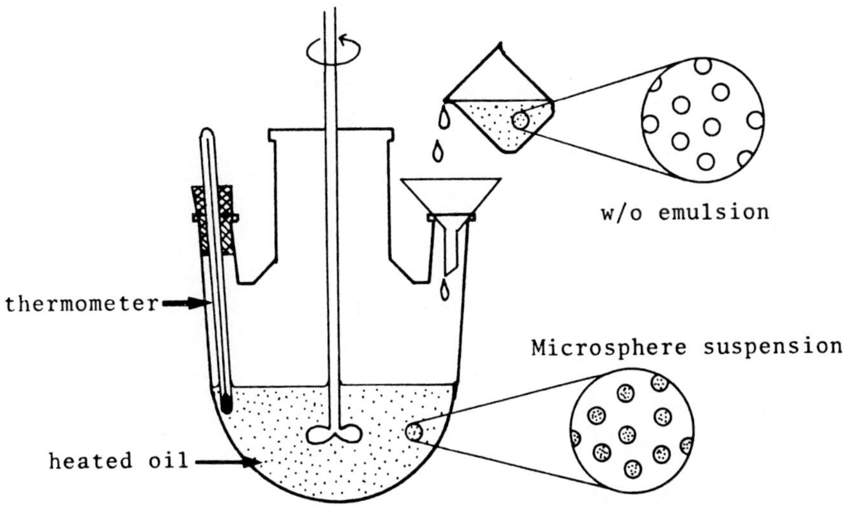

FIGURE 2. Apparatus for preparing albumin microspheres.

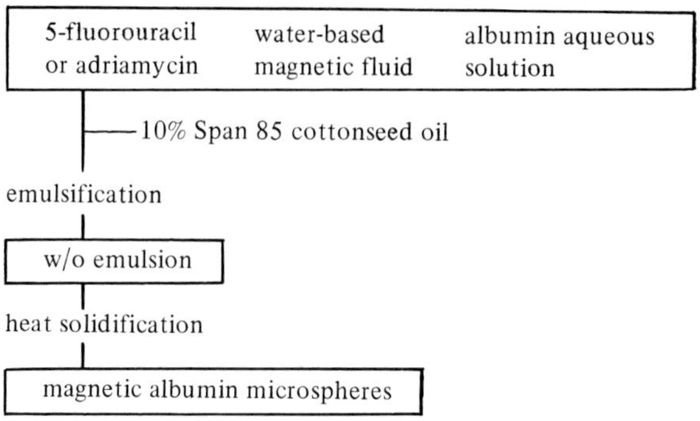

CHART 2. Schematic diagram of preparation of magnetic albumin microspheres containing 5-fluorouracil or adriamycin.

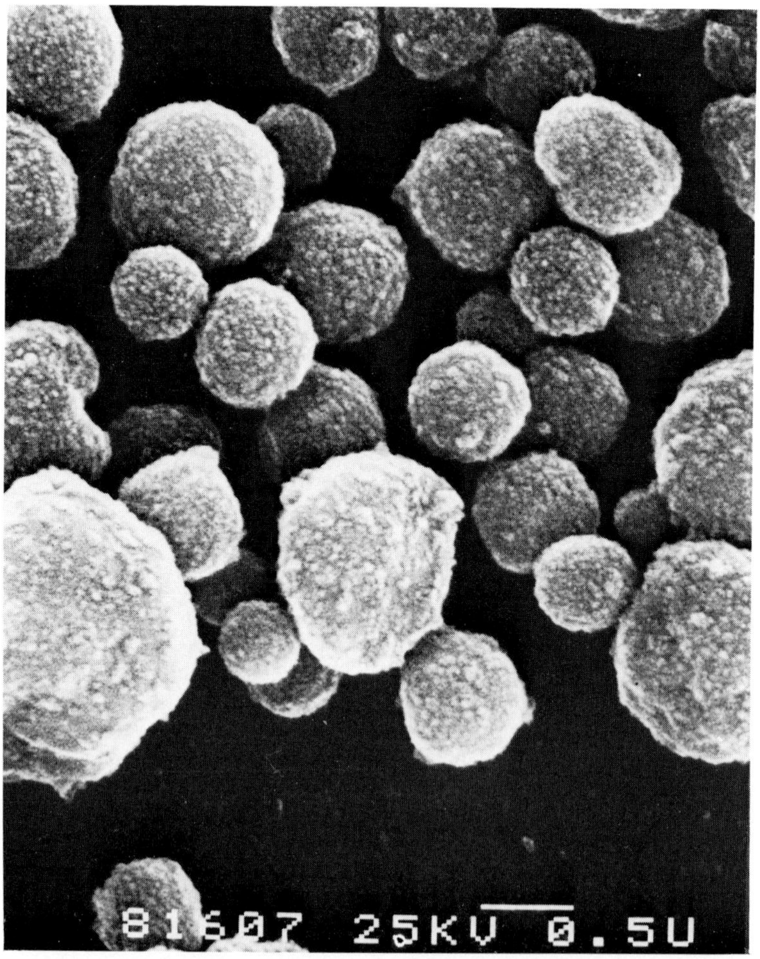

FIGURE 3. Scanning electron micrograph of magnetic microspheres.

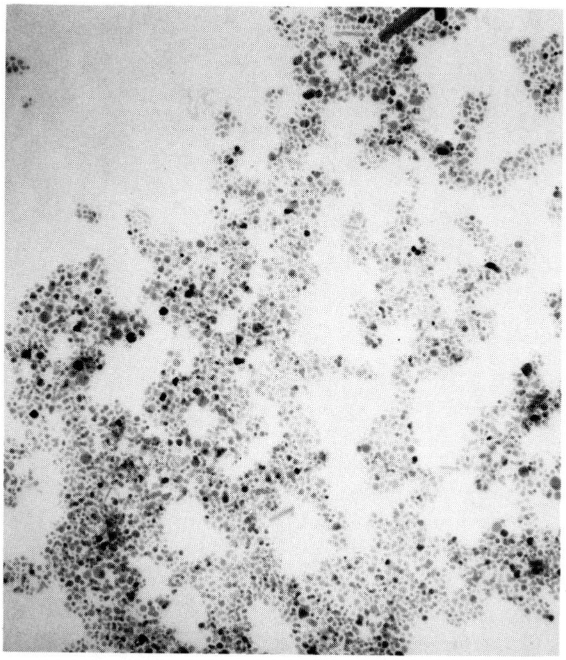

FIGURE 4. Transmission electron micrograph of water-based magnetic fluid.

ficult to deny the antigenesity of albumin microspheres. But it was suggested that the immunoreaction of the microspheres was very low compared with macroaggregate albumin and ferric oxide gel which were used as diagnostic agents in man.[36] From many studies of albumin microspheres as a scanning agent of the lung or liver in man, antigen-antibody reaction and/or hypersensitive reaction did not occur.[37]

We cannot say much about the toxicity of ferrite, because there are few published studies describing the toxicity of intravascularly administered Fe_3O_4. But studies performed on humans to evaluate the effects of the inhaled ferrite did not cause significant inflammation, fibrosis, or alteration of pulmonary function.[38-40] In addition, some Japanese researchers checked the toxicity of ferrite in mice and rats after oral administration and vascular-injection and did not find clear toxicity due to ferrite.[41-42] Based on these results, it is reasonable to predict that intravascularly administered microspheres will also induce minimal inflammatory responses. Since accumulation and removal of systemically administered ferrite are not well understood, the clearance and catabolism must be further examined in the near future. The putative biological effects, both toxic and beneficial, or high-intensity magnetic field represent a still controversial area. Although there are some problems, the magnetic albumin microspheres and the magnet used in the experiments might be generally safe.

3. *Magnetic Characterization*

Figure 5 shows sequential photographs of magnetic microsphere suspension in a petri dish. After a disc magnet was applied (3000 G), the magnetic microspheres started to move onto the magnet site. No more movement of microspheres could be seen after 60 sec. These results suggested that the magnetic albumin microspheres would be good candidates for magnetic drug carriers which rapidly respond to magnetic fields.

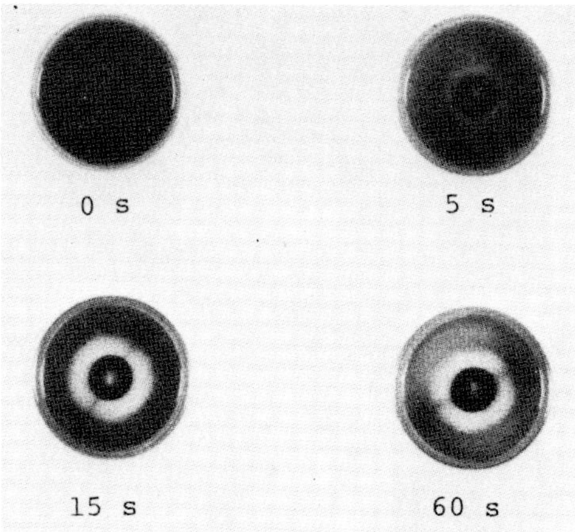

FIGURE 5. Time sequence photographs of magnetic microspheres in aqueous suspension after applied disc magnet.

4. In Vitro Drug Release

We measured the in vitro release phenomena of 5-fluorouracil (Figure 6) and adriamycin (Figure 7) from magnetic microspheres. 5-Fluorouracil release showed two processes; one is the initial burst release and the other is the next sustained release. In contrast, adriamycin release did not show the burst effect and the release rate is slower than that from 5-fluorouracil microspheres. Figure 8 shows a promotion of drug release from the microspheres using protease which digests many proteins. Since lysosome in cells has a high amount of enzymes, more active drugs may be released from the microspheres after the drug carriers are taken up into cells and digested with the lysosomal enzymes.[8]

C. In Vivo Magnetic Response of Magnetic Microspheres

1. Tissue Distribution after Intravenous Injection

In the first experiment, the mouse lung was selected as a model for in vivo testing of microspheres for two reasons: (1) the lung tumor occupies high ratios in many malignancies and (2) the microspheres after i.v. injection pass through the lung until sequestration in the liver. Each mouse was fixed on its back without anesthesia. One mg of the 1 or 3 μm microspheres was injected into the tail vein of mice in 0.2 mℓ of suspension. We prepared ^{125}I-labeled magnetic albumin microspheres using a tracer amount of ^{125}I-human serum albumin for the in vivo tissue distribution experiments. Two permanent magnets (Super Disc Magnet, No 30370, inner radius 5 mm, outer radius 9.5 mm, thickness 6 mm, Edmund Scientific) with a magnetic introduction of about 3000 G were directly applied to both sides of mice (the breast and back) throughout the experiment to concentrate the microspheres into the lungs. The mice were killed at 10 or 60 min after administration and ^{125}I-labeled microspheres in various tissues were determined by an auto gamma scintillation spectrometer (Packard®, Type 5110). The distribution of ^{125}I-albumin microspheres to various organs presented as percentage of dose per gram of tissue or that in the whole tissue.

Figure 9 shows tissue distribution of radioactivity at 10 min after i.v. injection of magnetic microspheres. After injection of the 1-μm microspheres in mice without magnets (control), about 3.9% of the administered dose (15.8%/g tissue at the concentration) were found in the lungs. When the particle size of the microspheres was enlarged from 1 to 3 μm in diameter, uptake of the 3-μm microspheres in the lungs increased from 3.9% of dose (15.8%/g tissue) to 10.7% (51.7%/g). When two magnets were

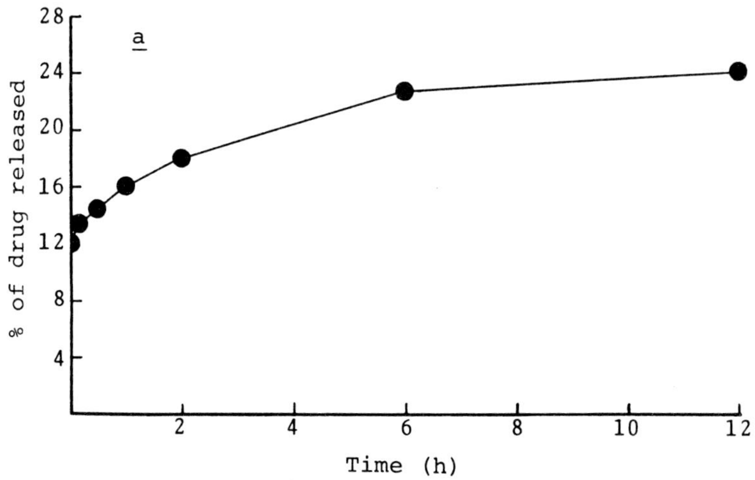

FIGURE 6. In vitro release of 5-fluorouracil from magnetic microspheres. Each point represents the average of three experiments.

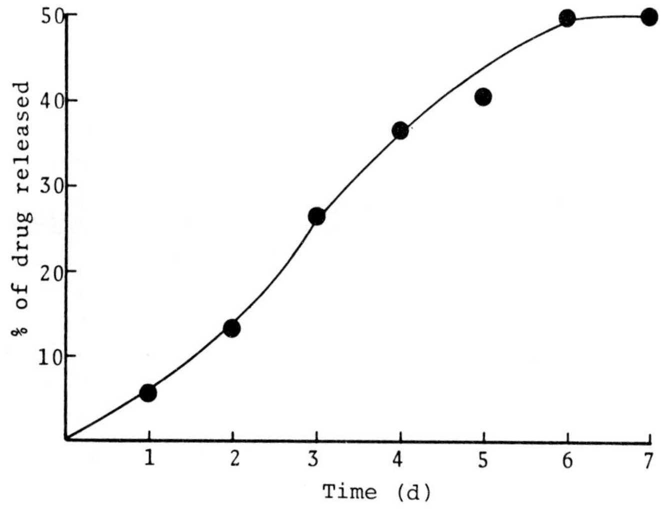

FIGURE 7. In vitro release of adriamycin from magnetic microspheres. Each point represents the average of three experiments.

applied to the lungs, the microsphere levels in the lungs increased about fourfold for the 1-μm microspheres and twofold for the 3-μm microspheres compared with each control. A peak lung level (104.0%/g tissue) was measured after injection of the 3-μm microspheres, but the amount of the microspheres in the lungs was only 21.6% of the dose.

Tissue distribution of radioactivity at 60 min after injection of the 3-μm microspheres in mice is shown in Figure 10, where there is also the distribution when two magnets were applied to the lungs only for the initial 10 min. The uptake of the microspheres in the lungs increased from 17.0% of the dose (67.2%/g tissue) to 28.2% (111.3%/g) and that in the spleen and liver decreased when two magnets were applied for 60 min. With application of two magnets for the first 10 min, there was an initial distribution of radioactivity to the lungs (Figure 9) followed by rapid clearance from the lungs and localization in the liver (Figure 10).

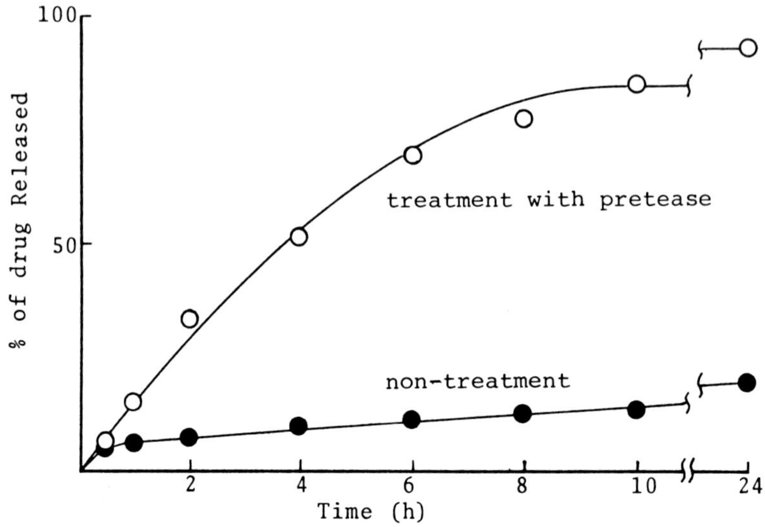

FIGURE 8. Effect of protease on the in vitro release of adriamycin from magnetic microspheres. Each point represents the average of three experiments.

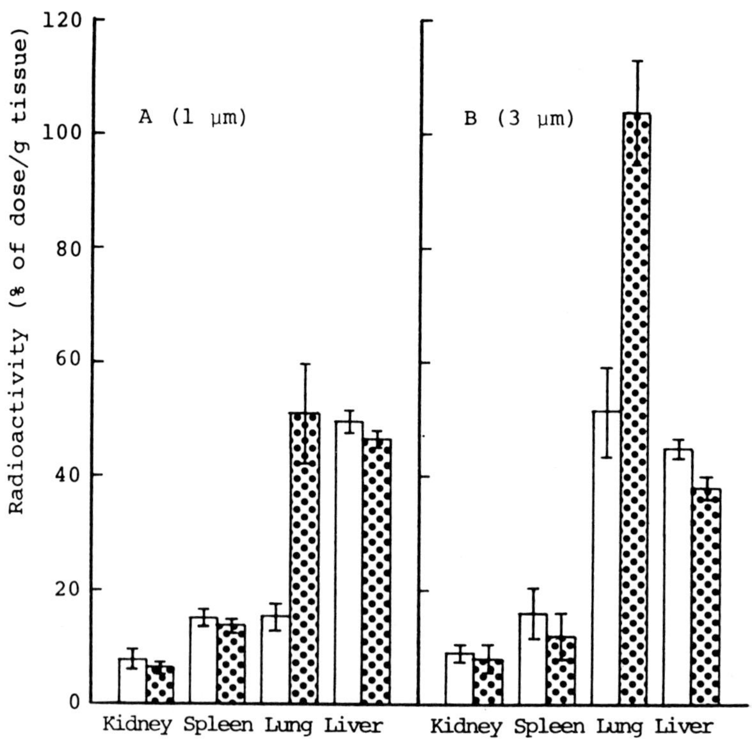

FIGURE 9. Tissue distribution of radioactivity at 10 min after i.v. injection of magnetic microspheres (1 and 3 μm in diameter) in mice: A, control (no magnet); B, treatment with two magnets throughout the experiments. Each column represents the mean ± SE of 3 to 5 mice.

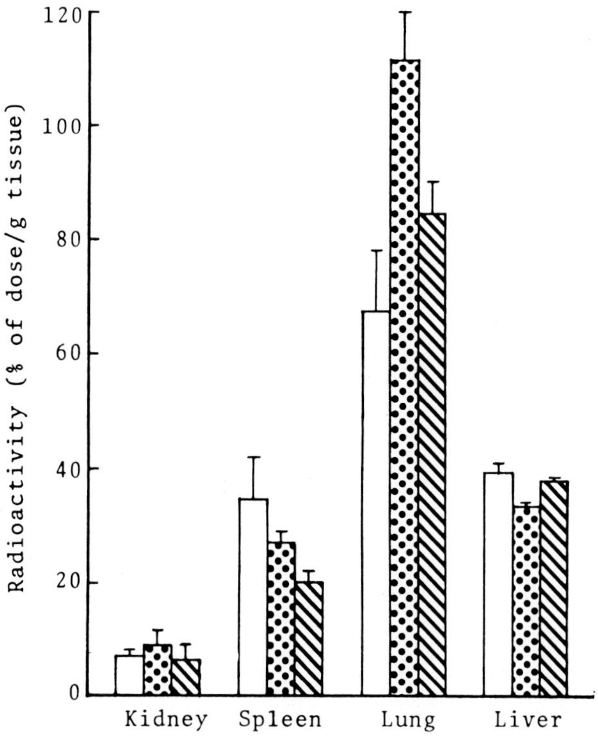

FIGURE 10. Tissue distribution of radioactivity at 60 min after i.v. injection of the 3 μm-magnetic microspheres in mice: □, control; ⊠, magnet (10 min); ⧅, magnet (10 min). Each column represents the mean ± SE of 3 to 5 mice.

2. Tissue Distribution after Intra-Arterial Injection

The kidney in rats was selected as a model in the second experiment. It is difficult to collect the microspheres in tissues beside the lungs or heart after i.v. injection, so the route of administration was changed to intra-arterial injection. Each rat was anesthetized with pentobarbital sodium at a dose of 60 mg/kg i.p. and was fixed on its back. Administration of the microspheres through the renal artery was carried out by cannulating a polyethylene tubing and forming a new vessel by-passing the original artery, because polyethylene tubing is easy to use for an injection. Into the left renal artery, 1 mg of the 1-μm microspheres was injected. Magnets (two) were directly placed throughout the experiments at both sides of the left kidney which was exposed by a midline abdominal incision. Tissue distributions of microspheres at 10 and 60 min after administration were examined by the method described above.

Figure 11 compares the tissue distribution of the 1-μm microspheres at 10 min following i.v. and intrarenal arterial administrations with or without application of two magnets to the left kidney. After i.v. injection, the microspheres did not concentrate in the left kidney and localized mainly in the liver regardless of application of two magnets to the kidney. The microsphere level in the kidney after intra-arterial administration was higher than that after i.v. injection with and without magnets. Administration into the rat's renal artery with magnets, on the other hand, concentrated the microspheres on the left kidney at 10 min to 56.4% of the dose (47.7%/g tissue) and the value was about 2.5-fold higher than that in the control.

Figure 12 compares the tissue distribution of the 1-μm microspheres at 60 min following intrarenal-arterial administration with application of two magnets to the left

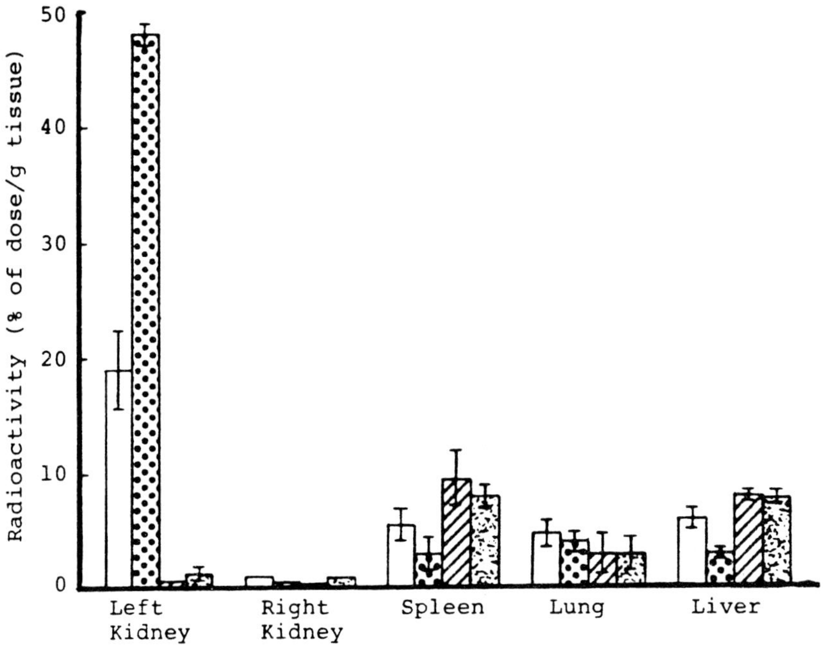

FIGURE 11. Tissue distribution of radioactivity at 10 min after i.v. or intra-arterial administration of 1 μm microspheres in rats: □, intrarenal artery (no magnet); ⊠, intrarenal artery (with magnet); ▨, intratail vein (no magnet); ▩, intratail vein (with magnet). Each column represents the mean ± SE of 3 to 5 rats.

kidney for the initial 10 or 60 min. Microsphere level in the left kidney was found to be 56.4% of the dose (47.7%/g tissue) at 10 min after administration, when applying the two magnets. In contrast to this value, the fraction of the dose remaining at the same organ at 60 min was much less when the magnet was removed 10 min after administration.

Recently, in several laboratories, it was realized that magnetic microspheres with magnetic responsiveness may be used as target-selective devices in cancer chemotherapy.[15,16,43] It is, however, experimentally obscure whether magnetic guidance of the microspheres to the target sites such as the lung and kidney is possible. In addition, no information has been collected regarding the effect of application time of the magnets to the target sites and removal from the sites on the behavior of the microspheres in blood. From these studies, it is evident that magnetic guidance of the microspheres to the sites, such as the lung and kidney, and the retention of the microspheres in those sites are possible if the administration route of the microspheres is carefully selected and application time of magnets to the target sites is extended (Figures 11 and 12).

A small fraction of the intravenously administered dose distributed to the lungs is due to a simple filter effect of the pulmonary capillary beds, although vascular constriction may play some roles in the trapping process. A large fraction is distributed mainly in the liver due to the reticuloendothelial cell activity. This phenomenon of filtration on the lungs was clearly observed when the particle size of the microspheres changed from 1 μm in diameter to 3 μm. However, with the 1- or 3-μm microspheres injected in mice, there was a large distribution to the lungs by application of the magnets, compared to no application of the magnets (Figure 9). Distribution of the microspheres to the kidneys following the intrarenal-arterial administration due to a simple filtration of the superficial and juxtamedullary glomerular layers[44] was not neg-

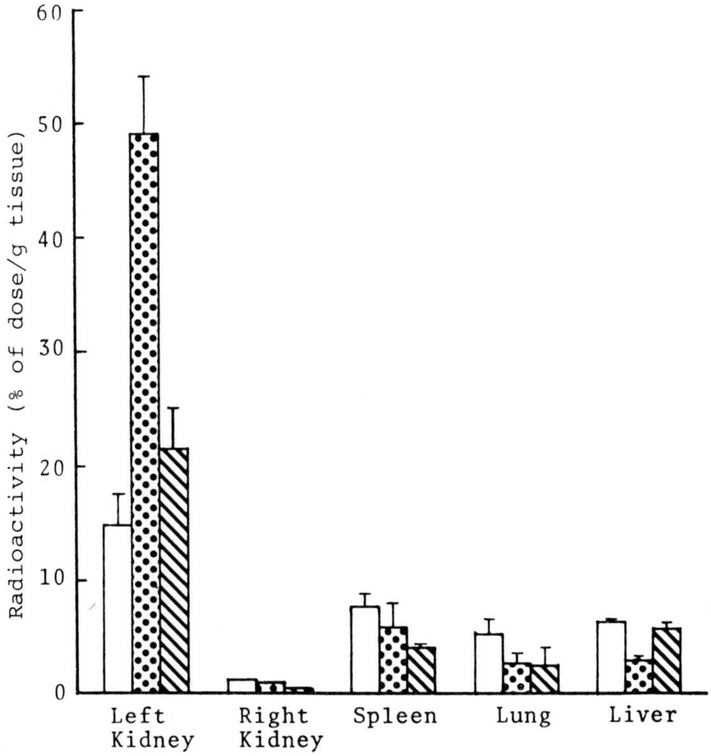

FIGURE 12. Tissue distribution of radioactivity at 60 min after intrarenal arterial administration of 1 μm-magnetic microspheres in rats: □, control; ▩, magnet (60 min); ▧, magnet (10 min).

ligible but the microsphere level in the kidney with the magnets was higher than that without the magnets (Figure 11).

With the 3-μm microspheres injected in mice, retention in the lungs when applying the magnets was greater than that without application. Those microsphere levels retained by an externally applied magnetic field in the lungs clearly decreased when the magnets were removed (Figure 10). The pattern of distribution in the kidney when the magnets were removed was similar to that in the lung (Figure 12). The decrease in the lungs or kidney between 10 and 60 min following administration of the microspheres may be due to wash-out within the lungs or kidney so that the microspheres initially trapped are allowed to recirculate in the bloodstream. The results obtained from the experiments of removing the magnets indicate that the application time of the magnets to the target sites is important for retention of microspheres in those sites.

A guarded selection of the administration route is also important for transporting microspheres to target sites. Kanke et al.[45] reported that no clear-cut difference in distribution patterns was observed between i.v. and intra-arterial administration of microspheres. Also Sjoholm and Edman[46] reported that the route of administration had no effect on the gross distribution pattern by studying the distribution of polyacrylamide microparticles after i.v. and i.p. injection in mice. The results shown in our experiments, however, found a clear-cut difference in the distribution pattern present between i.v. and intra-arterial administration of microspheres in rats. Thus, the in vivo distribution of the microspheres after different routes of administration is really complicated and further experiments and sufficient discussion would be needed.

The amount of microspheres retained at the target site (lungs in mice) was 28.2% of the dose at 60 min after i.v. injection of the 3-μm microspheres at a field strength of about 3000 G. Failure to achieve greater retention of microspheres is probably due to weak magnetic strength. When the strength of magnetic field is increased, this value might be improved considering that Widder et al.[16,47] used high magnetic field strength (8000 Oe) to retain 37 to 65% microspheres into the rat tail after infusion through the ventral caudal artery. As mentioned above, many drug carriers which are capable of transporting drug molecules from the site of application directly to the site of action have been developed.[48,49] In the near future some carriers of drugs may be used in the clinical field, but the utility of some drug carriers is restricted because of the lack of specificity or selectivity. Controlled localization of drug carriers, however, has been difficult to achieve. Recently, Gregoriadis[50] documented that attempts have been made to rationalize liposome development by tailoring their structure to the particular biological milieu in which they are intended to act. In the case of magnetic albumin microspheres, it is certain that this drug carrier can be used as a target-selective device.

D. Antitumor Effect of Magnetic Microspheres

1. Establishment of Experimental Methods for Evaluating Antitumor Effects

AH 7974 cells were selected as a model tumor. The tumor is an ascites hepatoma which was established in Japanese albino rats injected with *p*-dimethylaminoazobenzene in 1952 by investigators at the Sasaki Institute, Tokyo, Japan.[51] The tumor strain has been used in the screening of antitumor agents as well as the ascites forms of the Yoshida rat sarcoma, AH 130, AH 13, and AH 66 in Japan.

Adriamycin is one of the most effective antitumor agents against AH 7974 and the effect is higher than that with 5-fluorouracil, bleomycin, or cyclophosphamide.[52] In this experiment, AH 7974 cells were maintained by weekly transplantation in ascites form in Donryu rats. For preparing lung metastasis, AH 7974 cells (3×10^7 cells/0.3 mℓ) were implanted from the tail vein of rats.

The pole diameter, gap distance, and magnetic introduction of electromagnet (Nihon Denji Sokki, Japan) used to guide the magnetic microspheres in rats were 25 mm, 0 to 5 cm, and 6000/G at 10 V and 10 Å, respectively. Each rat was anesthetized with pentobarbital sodium at a dose of 60 mg/kg i.p. and both poles of the magnet were directly applied to the sides of each rat, (the breast and back) for 10 min after administration of the microspheres to concentrate the microspheres onto the lungs. The magnetic introduction which was developed on the thorax of the rats was adjusted to 6000 G. The gauss meter (Yokokawa, Type 3251, Japan) was used to check the magnetic introduction of the electromagnet in the animal experiments. Microspheres (5 mg), containing 300 μg of adriamycin were injected into the tail vein in 0.5 mℓ of 0.9% NaCl suspension containing 0.2% polysorbate 80 on 1, 8, and 15 days after inoculation of AH 7974 cells. The antitumor effect of free or entrapped adriamycin against AH 7974 lung metastasis was studied by change in lung weight, histological examination, and animal survival data. Only for the survival experiments, the nonentrapped magnetic microspheres as well as free or entrapped adriamycin were examined and the effect of the magnet (nonadministration) was determined.

2. Microsphere and Adriamycin Distributions in Rats

After i.v. injection of ^{125}I-magnetic microspheres containing ^{3}H (G)-adriamycin in normal rats, the tissue distribution of ^{125}I- and ^{3}H-radioactivities to various organs at 10 min was measured. Figure 13 shows the (a) ^{125}I- and (b) ^{3}H- radioactivities, that indicate microsphere levels and total adriamycin levels released from the microspheres plus entrapped in microspheres, respectively. The distribution of ^{3}H-radioactivity after injection of free adriamycin was also shown in Figure 13b. When the magnet was applied (magnet group), magnetic microsphere level in the lung was 19.7% of dose per gram tissue and the value was about two times higher than when the magnet was not

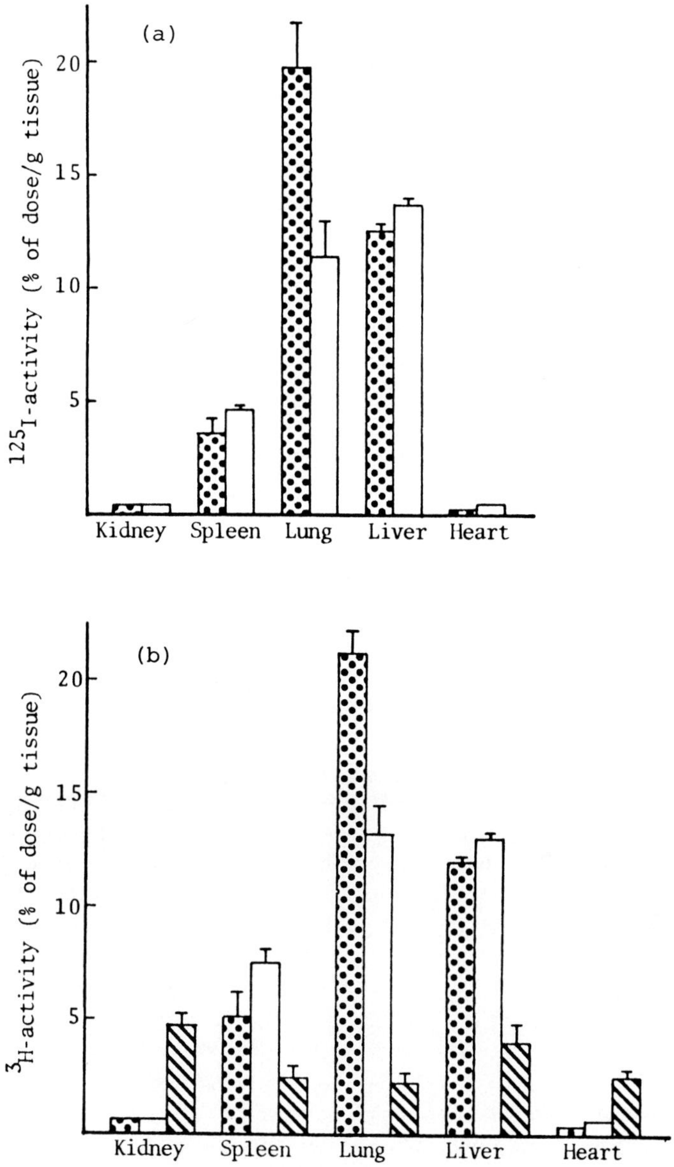

FIGURE 13. Tissue distribution of radioactivity after i.v. injection of ^{125}I-labeled magnetic microspheres containing ^{3}H-labeled adriamycin in rats: ▩, microspheres with magnet; □, microspheres without magnet; ▧, free adriamycin. Each column represents the mean ± SE of four rats.

applied (control group) (Figure 13a). The microsphere levels in the liver, spleen, and heart in magnet group were low compared with those in control group. The entrapped adriamycin level in the magnet group also increased and the value was eightfold higher than after injection of the free drug. The same results were obtained after injection of ^{125}I-magnetic microspheres into rats intravenously inoculated with AH 7974 cells (data were not shown).

3. Antitumor Effect of Adriamycin Entrapped in Magnetic Microspheres Against AH 7974 Lung Metastasis

First, the preventive effect against AH 7974 lung metastasis was evaluated by histological examination. Figure 14 shows microphotographs (HE stain) of the lung in (a) a normal rat, (b) a tumor rat on day 5, and (c) day 10 after inoculation with AH 7974 cells. Figure 15 shows microphotographs of the lung in the tumor rat treated with (A and B) free or (C and D) magnetic microsphere-entrapped adriamycin (A, C or B, D are photographs on days 5 or 10 after inoculation of tumor cells). AH 7974 cells were fixed into the lung tissues and grew locally day 5 after inoculation (Figure 14B). On day 10 (Figure 14C), the AH 7974 tumor grew further and the proliferation was marked around the blood vessels and trachea. In contrast, Figure 15A and B show that the tumor regions are thin compared with Figure 14B and C and that white blood cells infiltrate into lung tissues. These results suggested that free adriamycin affected the lung metastasis. After treatment with (C and D) magnetic microspheres containing adriamycin, the tumor cells are denaturated and the aveoli is clear compared with (A and B) that after treatment with the free drug and nontreatment (Figure 14B). These histological examinations showed that the magnetic albumin microspheres containing adriamycin exhibited higher effects against AH 7974 lung metastasis in rats than free drug.

Second, the antitumor effect of magnetic microspheres was evaluated by change in lung weight. Because AH 7974 lung metastasis spread not only around the blood vessels but also around the trachea and alveoli (Figure 14), the lung weight in tumor rats should increase rapidly. Figure 16 shows the change in lung weight in normal rats, magnetic microsphere-treated tumor rat, or nontreated tumor rats. The weight of the lung in normal rats increased slowly and average weight was about 0.7 g on day 0 (5 weeks old) and 1.3 g on day 15 (7 weeks old). The weight of the lung on day 15 after inoculation of AH 7974 cells was about 4.3 g, which was three times compared to that in normal rats. After treatment with magnetic microspheres, the average weight on day 15 was 2.4 g and the difference from that in normal rats was about 1.1 g. This value (1.1 g) was about one third of 3 g of the difference between the lung weights of nontreated tumor rats (4.3 g) and of normal rats (1.3 g). Thus, the magnetic microspheres containing adriamycin reduced in lung weight due to tumor growth.

Third, the antitumor effect was evaluated by animal survival data. Figure 17 shows the percent of rats in survival after inoculation of AH 7974 cells. Nontreated tumor rats (12 rats) died between 13 and 20 days after inoculation and median survival was 16.4 days. Data after nonentrapped magnetic microspheres (without adriamycin) were injected (five rats) are not shown. The median survival was 16.2 days, a value which was smaller to that of nontreated tumor rats (control rats). When treated with free adriamycin on days 1, 8, and 15 after inoculation (ten rats), the median survival increased to 21.2 days and T/C (median survival of treated rats/median survival of control, in percentage) was 129%. The value of survival time in rats treated with free drug was significantly different from that of control rats ($p<0.001$). In contrast, after magnetic microspheres containing adriamycin were injected into the rats applied with a magnet (ten rats), the median survival was 23.6 days and T/C was 144%, a value which was higher than in the free adriamycin group. The survival time in rats treated with magnetic microspheres was significantly different from that of control rats ($p<0.001$) and that of free drug group ($p<0.05$). Data after the microspheres containing drug were injected into rats without a magnet (5 rats) are not shown, where the median survival was 17.4 days. Survival time in rats treated with free adriamycin was higher than that when the rats were treated with microspheres containing adriamycin, applied with a magnet. The results might be explained by a difference of an effective adriamycin concentration in the target tumor sites (that is, not adriamycin level entrapped in the microspheres but free [nonentrapped] adriamycin).

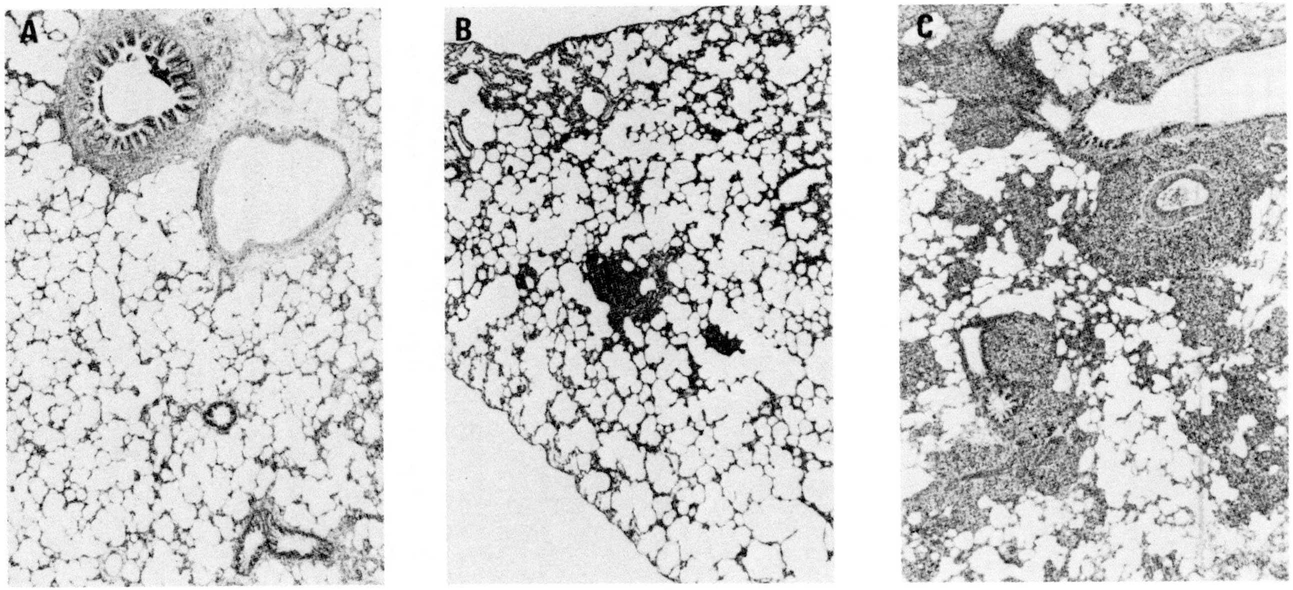

FIGURE 14. Photomicrographs of the normal and AH 7974 metastasized lungs: (A) normal lung; (B) tumor lung on day 5; or (C) day 10 after inoculation of AH 7974 cells (magnification × 100).

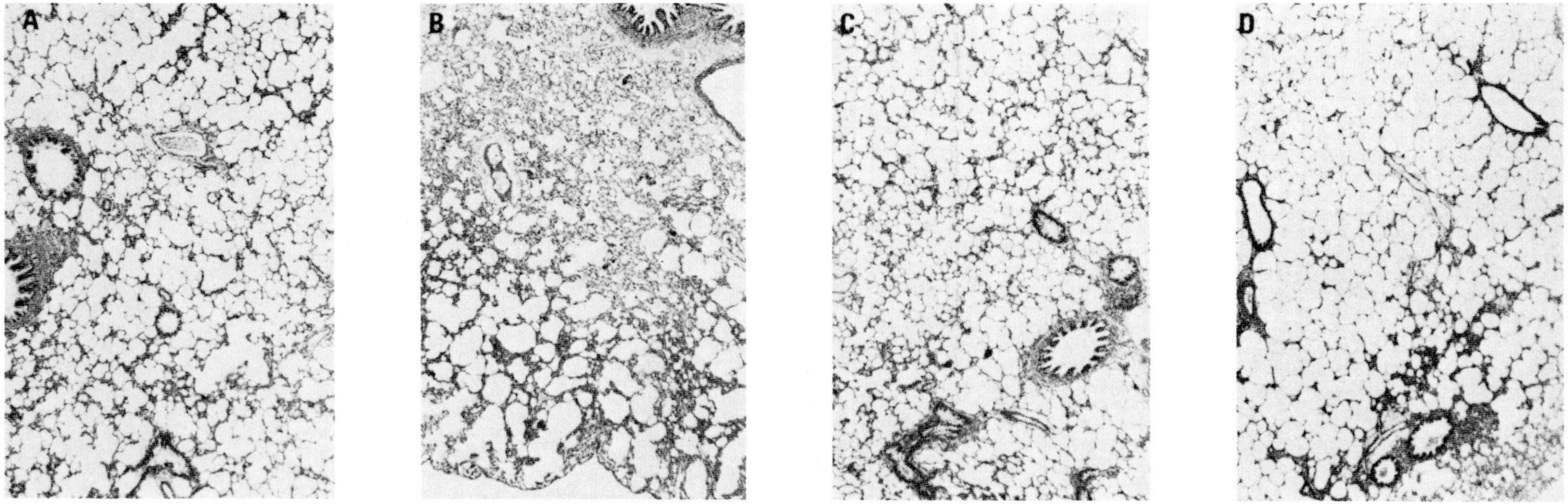

FIGURE 15. Effect of free or magnetic microsphere-entrapped adriamycin on AH 7974 metastasized lung. (A) Day 5 or (B) day 10 after injection of free adriamycin, (C) day 5 or (D) day 10 after injection of adriamycin-entrapped magnetic microspheres (magnification × 100).

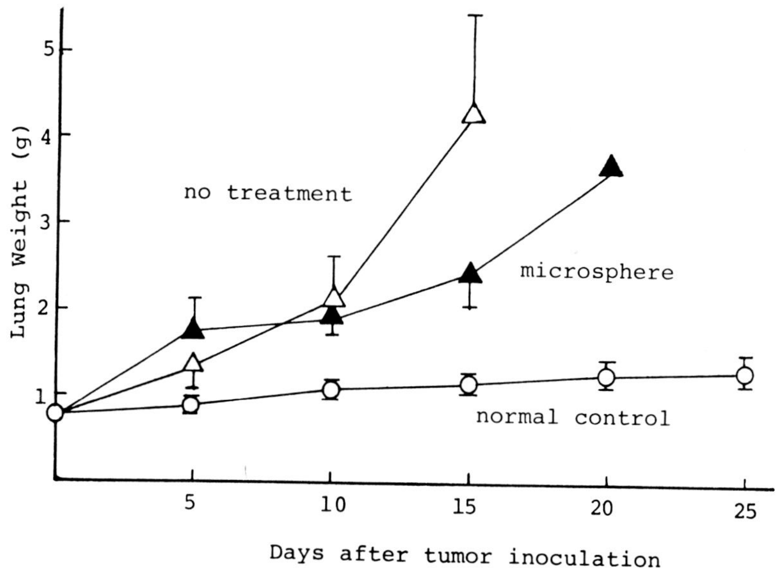

FIGURE 16. Change in lung weight of AH 7974 lung tumor-bearing rats. Each point represents the mean ± SE of five rats.

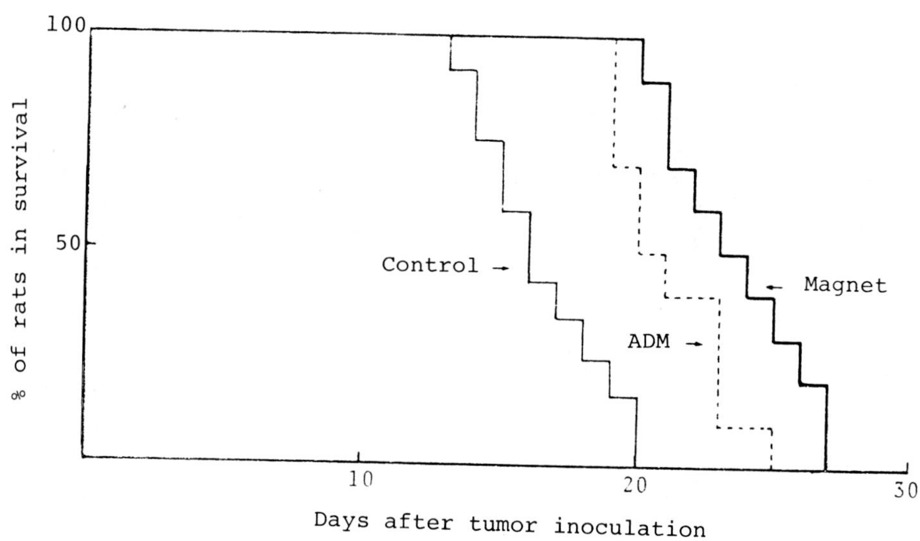

FIGURE 17. Effect of microsphere-entrapped adriamycin on the survival of rats inoculated with AH 7974 tumor cells.

E. Comment

Magnetic microspheres exhibited sustained drug release (Figures 6 and 7) and their albumin matrix would be biodegraded by hydrolytic enzymes such as protease (Figure 8). The microspheres could be guided into a target site in vivo (lungs, kidney) by magnetic means (Figures 9 to 13). The selective microsphere and entrapped drug distributions should be improved by the development of magnets suitable for the biological applications. Since the approaches to targeting of chemotherapeutic agents with previously reported drug carriers[8,53,54] were based on the physicochemical or biological interactions between the carriers and human bodies, clinical use of such drug carriers might be limited. The targeting with magnetic albumin microspheres, however, was due to the magnetic properties between the carrier and magnets and did not depend on the in vivo parameters. The availability of this targeted carrier system suggests potential uses in a number of localized disease other than tumors.

From histological examinations (Figures 14 and 15), change in lung weight (Figure 16) or animal survival data (Figure 17), enhancement of antitumor effects of adriamycin was exhibited by administration of magnetic microspheres containing adriamycin and subsequent exposure of a magnetic field to the lungs. This preventive effect against AH 7974 lung metastasis might be due to the selective microsphere distribution into a target site (lungs) by magnetic means and sustained release of adriamycin into the lungs from magnetic microspheres. Microsphere or adriamycin level in a target site and antitumor effect, however, were insufficient for an effective cancer chemotherapy. This problem might be solved to some extent by an enhancement of the power of applied magnets. If the problems can be overcome, selective cancer chemotherapy with antitumor agents then can be freely guided in vivo may be achieved by magnetically drug responsive albumin microspheres containing antitumor agents.

III. MAGNETIC EMULSIONS

Emulsion systems have been broadly used in pharmaceutical, fragrance, and food industry areas. Even in cancer chemotherapy, several kinds of emulsion formulations have been developed to find out sustained release and to improve the tissue distribution such as the lymphatic systems.[54] But targeting of emulsions except in the lymph is poor and stability of emulsions is not good, which have limited the usefulness of emulsion formulations in clinical stage.

This chapter describes the development of magnetically responsive oil-in-water type emulsions with the capacity to localize two antitumor chemotherapeutic agents, 1-(2-chloroethyl)-3-(trans-4-methylcyclohexyl)-1-nitrosourea (methyl CCNU) and 1-(4-amino-2-methyl-5-pyrimidinyl)-methyl-3-(2-chloroethyl)-3-nitrosourea (ACNU) by magnetic means to a specified target site. The magnetic emulsions consist of oleic acid- or ethyl oleate-based magnetic fluids as dispersed phases and casein solution as a continuous phase, with methyl-CCNU or ACNU being entrapped in the oily dispersed phase.

In addition, the antitumor activities of such chemotherapeutic agents entrapped in magnetic emulsions were examined against AH 7974 lung metastasis in rats.

A. Preparation of Magnetic Emulsions

Oil-based magnetic fluids used in these experiments were prepared with oleic acid or ethyl oleate. The synthesis method of magnetites was modified with a method of Shimoiizaka et al.[27] Oleic acid and ethyl oleate-based magnetic fluids contain 31.5 and 23.1% (w/w) magnetites in oleic acid and ethyl oleate, respectively.

The compositions of two oil-in-water (o/w) type magnetic emulsions are shown in Table 1. Casein was selected as a surfactant with preliminary stability studies of magnetic o/w emulsions using protein.[23] Proteins are worthy to notice as a neutral emulsifier. Casein at pH 8 derived the most stable point among the several combinations of proteins and pHs. The stability of magnetic emulsions may be reviewed in detail in Reference 23. For preparing magnetic microspheres, antitumor agent ACNU or methyl-CCNU, was dissolved or dispersed uniformly in oleic acid- or ethyl oleate-based magnetic fluids prior to the emulsification. Then, the magnetic fluids and 1% casein solution (pH 8) were mixed and agitated with a high speed stirrer (Ultra Turrax) at 10,000 rpm for 5 min. In all cases, emulsification was carried out in a water bath at 37°C. The diameter of oil drops in magnetic emulsions was 1 to 20 μm and the averages for oleic acid-based and ethyl oleate-based emulsions were 6.4 and 5.0 μm, respectively.

Table 1
COMPOSITION OF MAGNETIC EMULSION FORMULATIONS

Oleic acid-based magnetic emulsion	
Water phase	6.0 mℓ
Casein	60 mg
Water	q.s
Oil phase	1.5 mℓ
Magnetite	472 mg
Antitumor agent	45 mg
Oleic acid	q.s.
Total volume	7.5 mℓ
Ethyl oleate-based magnetic emulsions	
Water phase	6.0 mℓ
Casein	60 mg
Water	q.s.
Oil phase	1.5 mℓ
Magnetite	347 mg
Antitumor agent	45 mg
Ethyl oleate	q.s.
Total volume	7.5 mℓ

B. Several Properties of Magnetic Emulsions

1. In Vitro Magnetic Response

It is a difficult problem to trap and retain magnetic colloidal particles injected intravascularly with an externally applied magnetic field whenever magnetically responsive drug carriers are developed. Such carriers need enough magnetic responsiveness to gather in a predetermined magnetic site under the physiological conditions. In order to discuss this problem, trapping experiments of magnetic emulsions which flowed in a glass tubing were done.

An apparatus for measuring the magnetic responsiveness of the emulsions is shown in Figure 18. The blood flow in man varied from 0.2 to 1.0 mm/sec for small arteries to 30 to 40 cm/sec for big arteries and the diameter of blood vessels varied from 8 to 20 μm for capillaries to 3 cm for the pulmonary artery. We selected a glass tubing with an inner diameter of 0.6 cm and a laminar flow rate of 15 cm/sec as a model for middle blood vessels. The flow medium was an isotonic phosphate buffer (pH 7.4). A continuous-flow-rate pump (Type RF, Tokyo Rika Kikai, Co. Ltd., Japan) was used to establish laminar flow rate in glass-tubing (30 cm in length). The tubing was positioned between both poles of the electromagnet as shown before. Oleic acid- or ethyl oleate-based magnetic emulsions (0.5 mℓ) were injected into the introduction chamber made of silicone tubing after the whole tubing was filled with phosphate buffer. The emulsions retained by a given magnetic flux density (500 to 6000 G) were collected and digested with sulfuric acid and hydrochloric acid. A small aliquot of the solution was then analyzed for iron with an atomic absorption spectrometer (Type AA-8200, Nippon Jarrel-ASH Co. Ltd., Japan) equipped with a heated graphite atomizer. For another set of experiments, the magnetic emulsions containing ^{14}C-palmitic acid as a tracer in oil drops of emulsions were injected. The amount of oil phase could be measured for radioactivity by a liquid scintillation counter (Model LSC 703, Aloka, Japan). Each retained amount of emulsion was calculated as a percent of the total amount.

Results for only ethyl oleate based-magnetic emulsions are shown in Figure 19. About 30.8% of ethyl oleate (measured with ^{14}C-palmitic acid) and 33.0% of magnetite (measured with iron) per initial value were retained at the target site at 6000 G of magnetic flux density. As shown in this figure, the amounts of oil and iron retained in

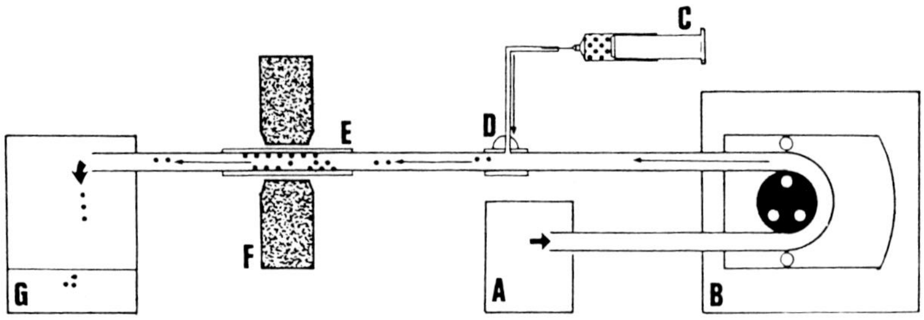

FIGURE 18. A constant-flow apparatus used to measure the magnetic responsiveness of magnetic emulsions.

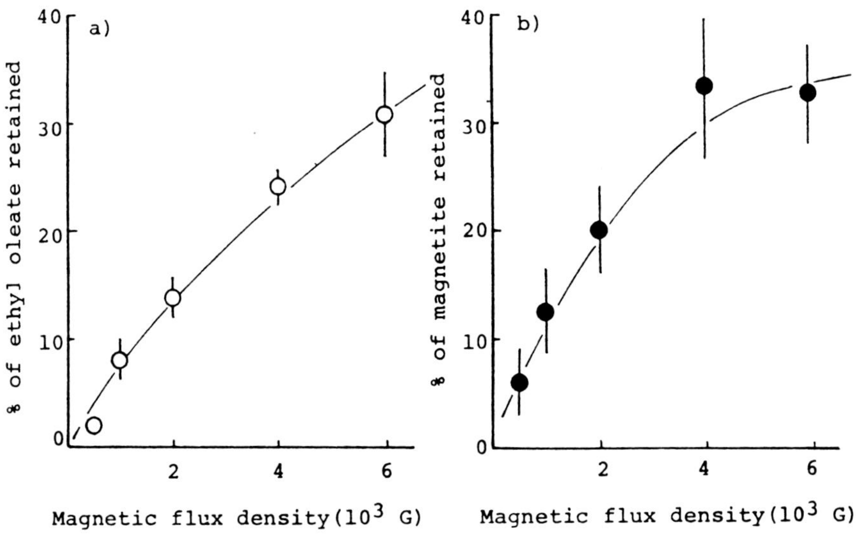

FIGURE 19. Retention of ethyl oleate-based magnetic emulsion flowing in isotonic buffer as a function of magnetic flux density. (a) Ethyl oleate; (b) magnetites.

magnetic sites show the same phenomenon, (i.e., the retention of the emulsions significantly increased at higher magnetic flux density). Similar data were found in oleic acid-based magnetic emulsions. The results suggested that the magnetic emulsions prepared with oleic acid- and ethyl oleate-based magnetic fluids could be retained in a magnetic field even if the medium was flowing as in the blood vessels. The amount retained may depend upon the stability of magnetic fluids, flow rate, magnetic power and/or pole area of magnets. Since we did not change the pole area in these experiments, a large amount of emulsions retained at a magnetic site of the inside of tubing was over 4000 G. This phenomenon may be a reason for the saturate level of emulsions retained as shown in Figure 19b. This experimental method was not enough for evaluating the in vitro magnetic response for magnetic drug carriers, but the results showed a qualitative character of magnetic emulsions.

2. Stability of Nitrosoureas

Magnetic o/w emulsions are useful carriers to entrap some lipophilic drugs such as ACNU and methyl-CCNU in the inner oil phase and to change the distribution character by an application of magnetic fields. Since lipophilic nitrosoureas, (i.e., BCNU) are generally unstable in aqueous solution, it has been difficult to make a sufficient parenteral formulation.

Our experiments regarding the stability in pH buffers and plasma[55] showed that 50% of methyl CCNU and ACNU decomposed within 8 to 60 min in pH 7 buffer and 20 to 30 min in rabbit plasma at 37°C, respectively. In addition to the rapid decomposition of these nitrosoureas, their decomposition compounds, isocyanate and 2-chloroethyl carbonium, have cytotoxicity. In contrast, nitrosoureas in ethyl oleate were more stable than that in aqueous solutions as shown in Table 2. These results suggested that methyl-CCNU and ACNU entrapped in oil-based emulsions might be stable and useful to minimize the side effects due to the degradation products.

3. *In Vitro Drug Release*

Nitrosoureas released from the emulsions were determined by an equilibrium dialysis system with a cellulose tubing at 37°C. The concentration of nitrosoureas, methyl-CCNU, and ACNU in the aqueous solution was determined colorimetrically by the method of Loo and Dion.[56]

The release patterns of methyl-CCNU and ACNU from the magnetic emulsions are shown in Figure 20. ACNU entrapped in magnetic emulsions released rapidly to the outer solutions in the early periods. On the other hand, the release of methyl-CCNU from the emulsions was much slower than that of ACNU. The difference of release patterns could be interpreted as the difference between the partition coefficients of the two drugs. The apparent total amounts of methyl-CCNU released into the buffer or saline were 65.7 or 157.8 μg, respectively. Similarly, the amounts of ACNU released into the buffer (pH 4.0) or saline were 1291.4 or 840.5 μg, respectively. Thus, it is expected that the drug release rate could be controlled by replacing entrapping drugs.

4. *In Vivo Magnetic Response*

Female Donryu rats, weighing about 140 to 150 g, were used in this experiment. The site utilized for in vivo targeting of the emulsions was the rat lung. Each rat was anesthetized and both poles of the electromagnet were applied to the lungs of each rat for 10 or 60 min after injection of the emulsions so as to collect the emulsions into the lungs (the same as experiments for magnetic microspheres). The value of magnetic introduction developed in the thorax in rats was adjusted to 2000, 4000, or 6000 g. For determining the tissue distribution of emulsions, magnetic emulsions containing ^{14}C-palmitic acid in the oily phase were used. The emulsions (0.2 mℓ) were injected into the rats through the tail vein. Following injection, the electromagnet was retained for 10 or 60 min and the rats were sacrificed by decapitation. For measurement of ^{14}C-levels, each aliquot of tissue was solubilized with Protosol® (New England Nuclear), diluted directly with a scintillator (Aquasol®, New England Nuclear), and determined with a liquid scintillation counter.

In order to make sure the responsiveness of magnetic emulsions in histological level, the lungs and liver were stained with Berlin Blue at 10 min after i.v. injection of magnetic emulsions.

The tissue distribution of radioactivity at 10 min after injection of magnetic emulsions is shown in Figure 21. After injection of emulsion into rats without electromagnets (control group), about 16.2% of magnetic emulsions injected were found in the lungs. When the electromagnet was applied to the lungs (magnetic group), the emulsion level in the lungs was 28.9, 32.5, or 32.7% at 2000, 4000, or 6000 G, respectively, i.e., about twice compared to the control group. However, there was not a clear correlation between the emulsion levels in the lungs and the magnetic flux density.

Figure 22 shows the tissue distribution of the emulsions at 60 min following i.v. injection with application of the electromagnet to the lungs for 60 min. Lung emulsion levels in the magnetic group were found to be about 10 to 11% of dose at 60 min after injection regardless of the intensity of magnetic induction. However, about 8.0% of

Table 2
DECOMPOSITION OF ACNU AND METHYL-CCNU IN ETHYL OLEATE AT 37°C

Time (day)	Percent of unchanged substance	
	ACNU	Methyl-CCNU
0	100.0	100.0
1	94.1 ± 0.2	78.8 ± 3.9
7	71.7 ± 10.1	64.1 ± 17.5

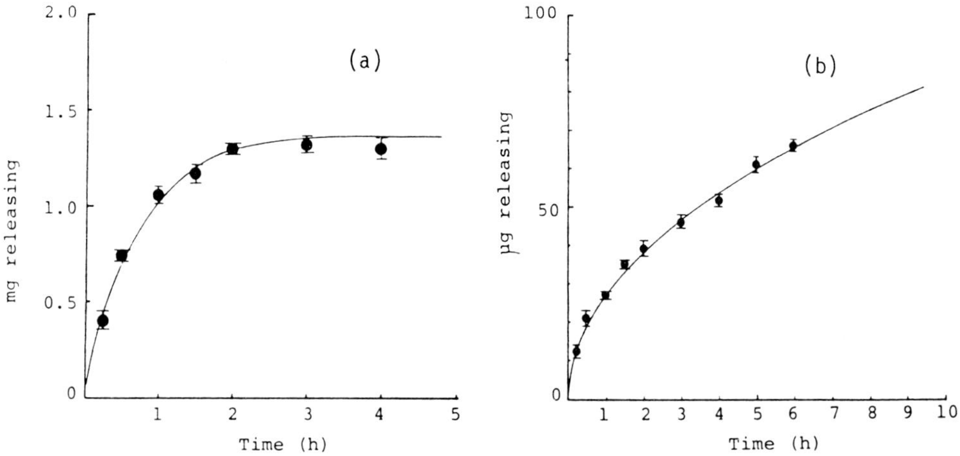

FIGURE 20. In vitro release of ACNU and methyl-CCNU from magnetic emulsion. (a) ACNU; (b) methyl-CCNU.

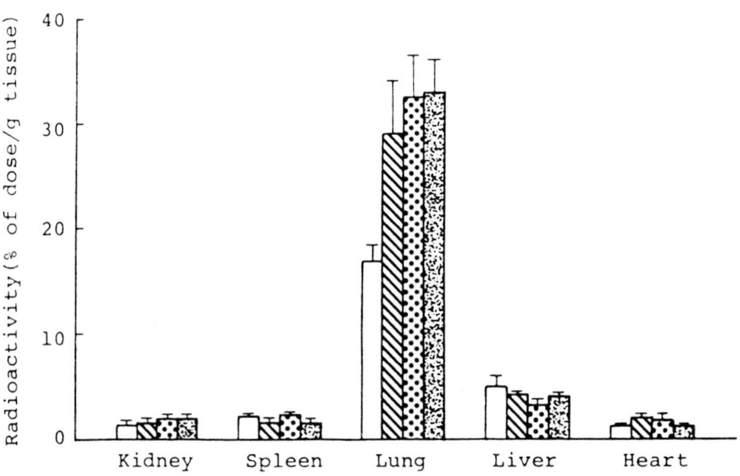

FIGURE 21. Tissue distribution of radioactivity at 10 min after injection of ethyl oleate-based magnetic emulsions containing ^{14}C-palmitic acid through the tail vein: □, control; ▧, magnet (2000 G), ▩, magnet (4000 G), ▨, magnet (6000 G). Each column represents the mean ± SE of five rats.

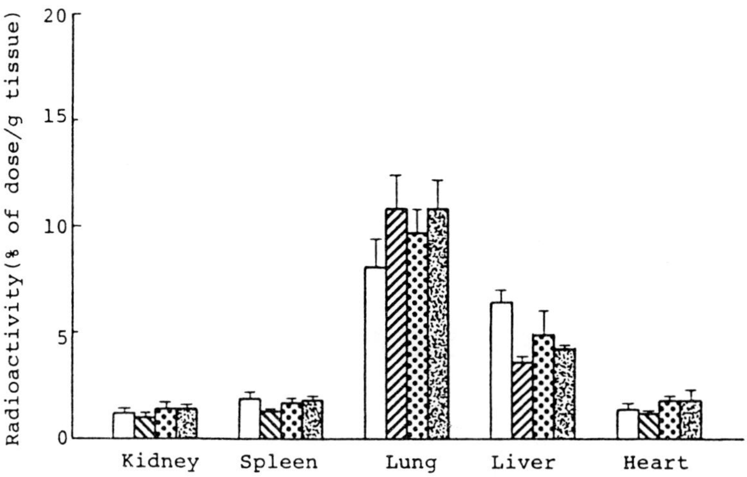

FIGURE 22. Tissue distribution of radioactivity at 60 min after injection of ethyl oleate-based magnetic emulsions containing^{14}C-palmitic acid through the tail vein: column; as in Figure 21. Each column represents the mean ± SE of five rats.

radioactivity was found in the lungs at 60 min after injection into rats without the electromagnet. The difference between the magnet and control groups was less than we had expected.

Figures 23 and 24 show photomicrographs of the lungs and liver 10 min after i.v. injection of magnetic emulsions, respectively. This study showed that magnetites contained in emulsions reached to the lungs, lodged in the pulmonary capillary beds, and/or were retained in the immediate vicinity of the alveoli. However, magnetites that passed the pulmonary filter were mainly trapped in the liver. The reason for the high amount of magnetic emulsions in the lungs might be explained by these results, but the difference between control and magnetic groups was not substantial.

5. Tissue Distribution of ACNU Entrapped in Magnetic Emulsions

Magnetic emulsions containing ACNU were injected into rats. The dose of ACNU was 10 mg/kg rat. The rats were sacrificed and the organs and blood were withdrawn. ACNU was determined by the modification of the ion-pairing HPLC method.[57]

Dichloromethane and 2-phenylbenzimidazole were used as an extraction solvent of ACNU from tissue homogenates in pH 7.4 phosphate buffer and as an internal standard in HPLC assay, respectively. Final samples (25 $\mu\ell$) were chromatographed using a stainless steel precolumn (30 mm × 4.6 mm, I.D.) and main column (250 mm × 4.6 mm, I.D.) of microporous silica bonded with octyl and octadecyl alkyl-groups (RP-8, Braunlee Lab. Inc., Santa Clara, Calif. and Nucleosil 5C18, Nergel, W. Germany, respectively). The eluting solvent was water-methanol (50:50) containing 0.1% sodium pentansulfonate and 1% (v/v) aceteic acid, at a flow rate of 1 mℓ/min. Detection was normally done by UV absorption at 244 nm. ACNU and internal standard were well resolved from the peaks of biological compositions and retention times of ACNU and internal standard under these conditions were 6 and 11 min, respectively.

Figure 25 shows the tissue levels 10 min after i.v. injections of aqueous solution or magnetic emulsion. It is clear that ACNU injected in the form of aqueous solution showed no specific localization and only a small amount was transferred to the target sites (lungs). ACNU levels in the lungs were less than 1 μg/g tissue at 10 min following injection and were quite low compared with that in other tissues except for the liver. In contrast, accumulation of the drug in the lungs when the electromagnet was applied

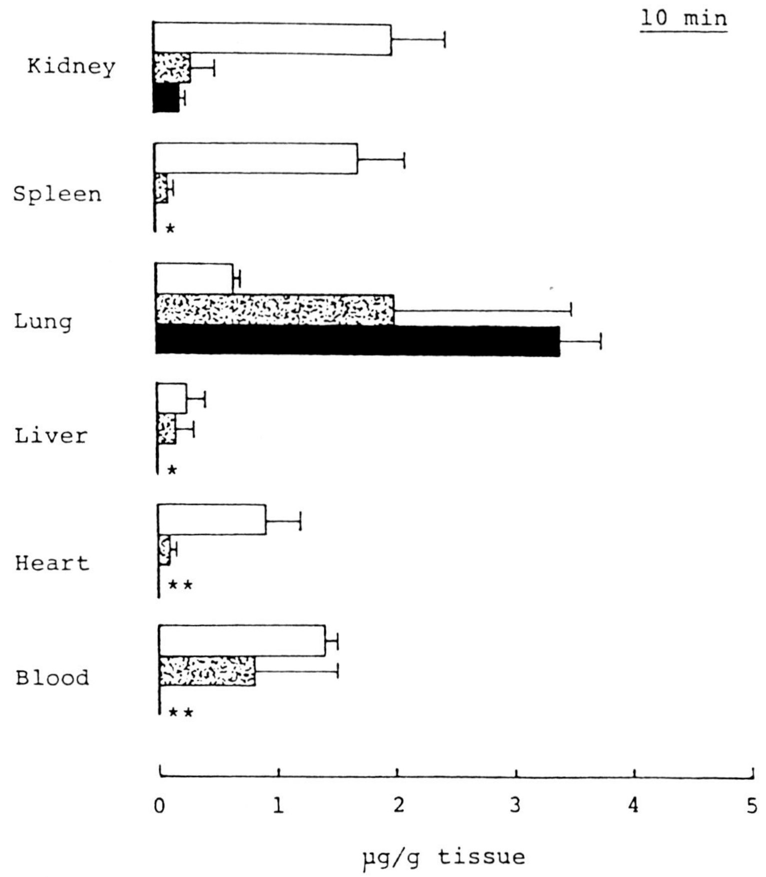

FIGURE 25. Amount of ACNU in various organs at 10 min after i.v. injection of magnetic emulsions with and without electromagnet. □, Aqueous solution; ▨, magnetic emulsions (no magnet); □, magnetic emulsions (with magnet); * not detected; ** trace level.

to the lungs (magnet group) was about five times higher than that when the drug was injected as an aqueous solution.

The drug concentration in each tissue 60 min following injection is shown in Figure 26. At 60 min postinjection of aqueous solution, unchanged ACNU was scarcely detectable in the lungs and kidney. Consequently, rapid decomposition and elimination from the body were estimated. At the same time, however, large and practically comparable amounts of the drug were present in the lungs after injection of emulsions into rats without the electromagnet. Moreover, when the electromagnet was applied to the lungs, ACNU still remained as high as 4.40 μg/g tissue in the lungs of animals that received the drug as magnetic emulsions. The amount of ACNU retained at a target site (lungs) in the magnet group was about four times higher than that in the control (nonmagnet) group.

At 6 hr after injection of the emulsions, the concentration of parent ACNU in the lungs was detectable for the magnet group (data were not shown). But ACNU was not consistently quantifiable in the lungs for control group. In magnet group, trace amounts were still demonstrable in the lungs even 24 hr after administration.

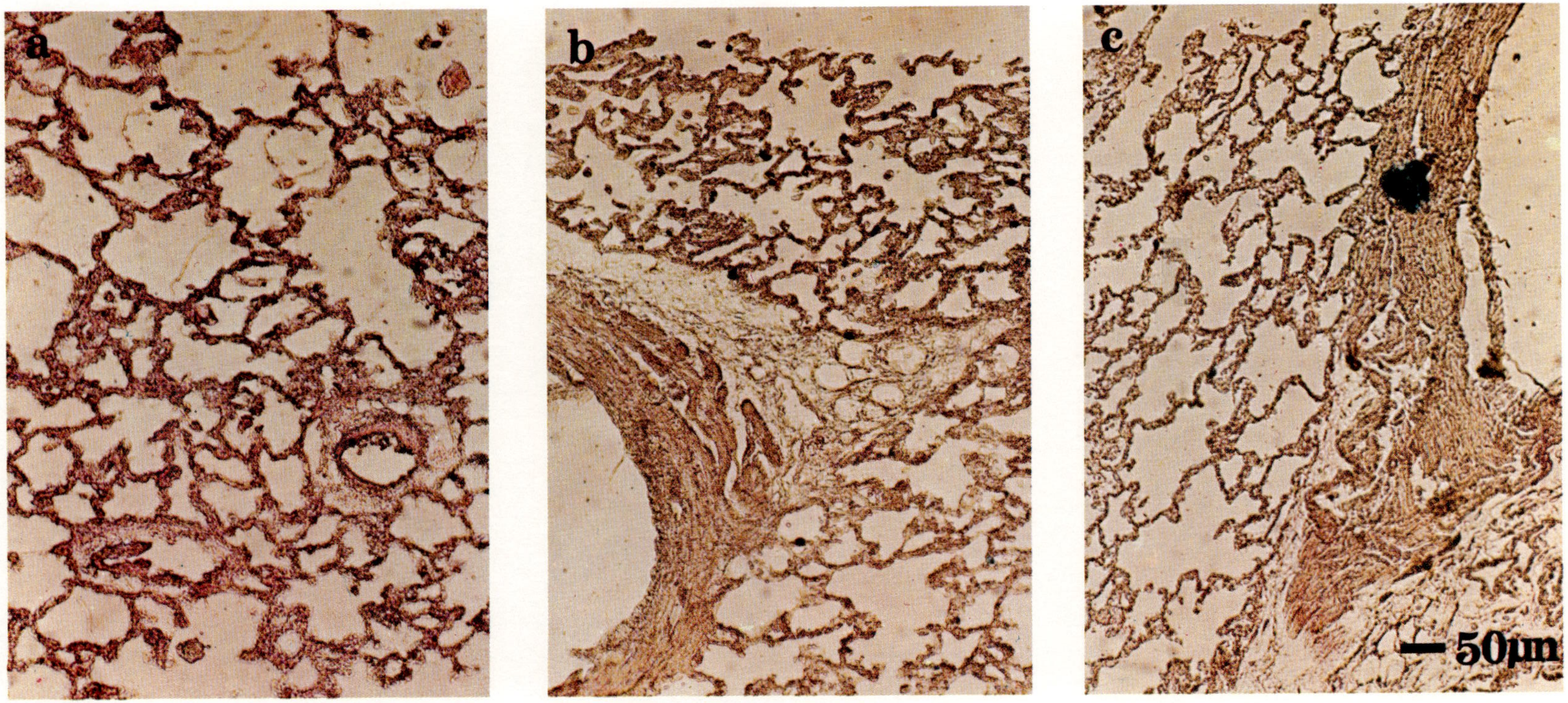

FIGURE 23. Photomicrographs of the lung (Fe stain) at 10 min after i.v. injection of magnetic emulsions. (a) Control; (b) nonmagnet; (c) magnet.

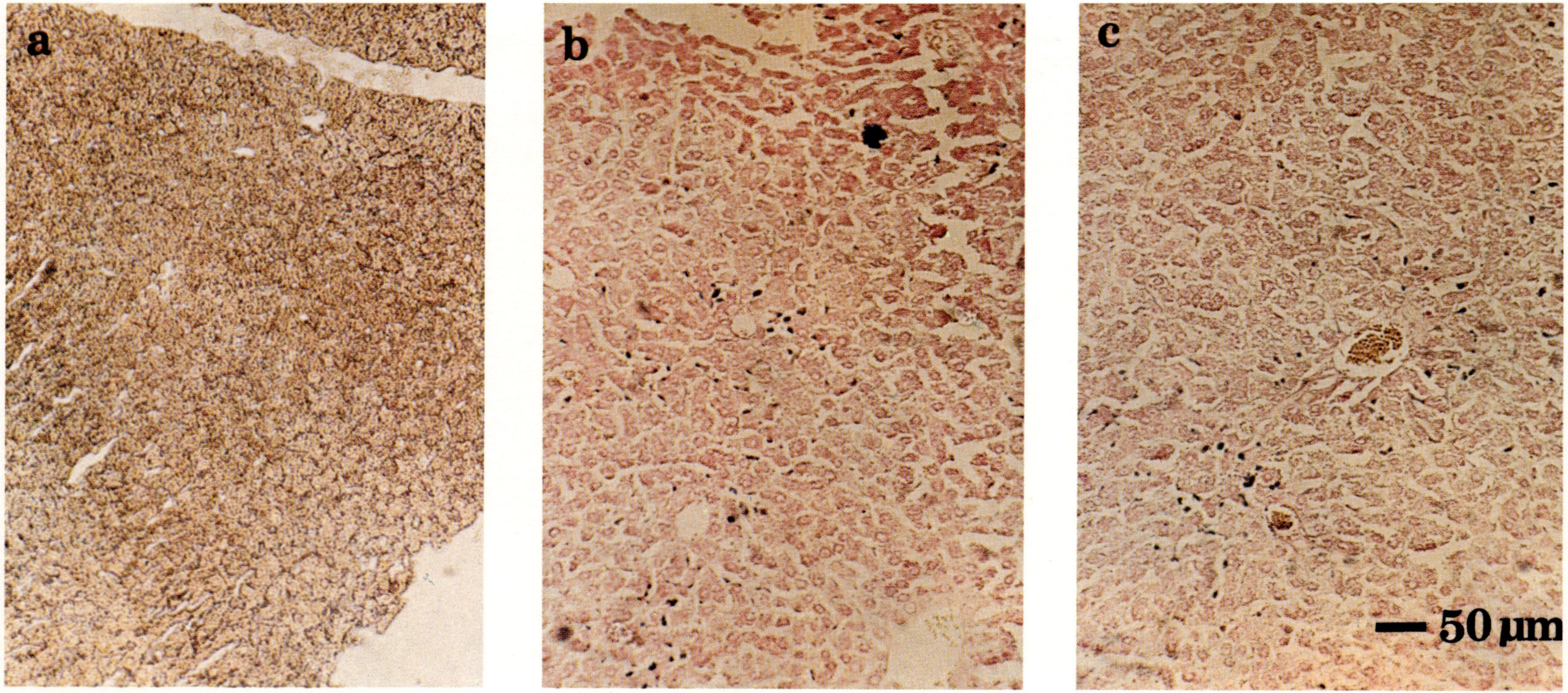

FIGURE 24. Photomicrographs of the liver (Fe stain) at 10 min after i.v. injection of magnetic emulsions. (a) Control; (b) nonmagnet; (c) magnet.

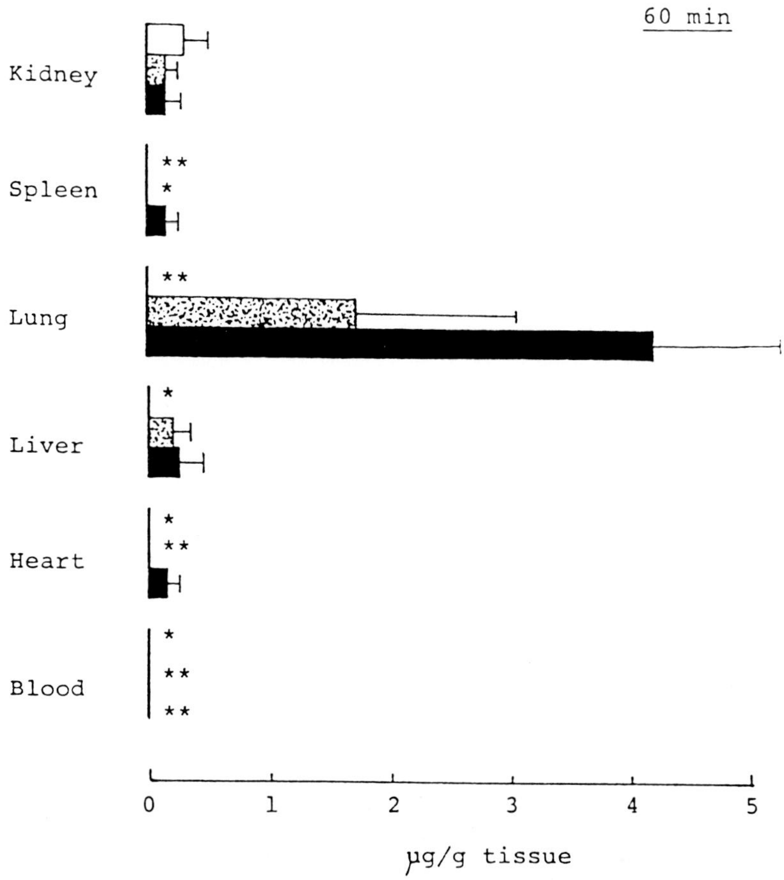

FIGURE 26. Amount of ACNU in various organs at 60 min after i.v. injection of magnetic emulsions with and without electromagnet. Column and captions as in Figure 25.

C. Antitumor Effect of Magnetic Emulsions

1. Antitumor Effect of Methyl-CCNU Entrapped in Magnetic Emulsions Against AH 7974 Lung Metastasis

The antitumor effect was evaluated on the basis of animal survival data. Table 3 shows the effect of methyl-CCNU administered with a single dosage of 10 mg/kg body weight on the survival time of Donryu rats born with AH 7974 lung metastasis. The untreated tumor bearing rats (control group) died between 14 and 22 days after inoculation and mean survival was 17.6 days. Intraperitoneal injection of methyl-CCNU as a suspension in 0.2% polysorbate 80 saline did not produce much difference in the length of survival from that of control group. When magnetic emulsions containing methyl-CCNU were injected intravenously into rats without application of an electromagnet, the average survival time increased to 19.9 days and T/C was 113%. Furthermore, after magnetic emulsions containing methyl-CCNU were injected into rats with 6000 G of magnetic induction to the lungs for 10 min, the mean survival increased to 22.9 days and T/C was 130%. The value of survival time in rats treated with the magnetically targeted emulsions containing methyl-CCNU was significantly different from that of the control group ($p < 0.01$) and that of rats receiving drug-entrapping magnetic emulsions without magnets ($p < 0.05$). The results might be explained by a difference of an effective drug concentration in the target cancerous lesions.

Table 3
EFFECT OF VARIOUS FORMULATIONS OF METHYL-CCNU ON SURVIVAL TIME OF RATS BORN WITH AH 7974 LUNG CARCINOMA

Formulations	Number of rats	Mean survival time ± SE (days)	T/C%[a]
Control[b]	16	17.6 ± 2.6	100
Methyl-CCNU-suspension[c]	8	17.3 ± 2.3	98
Methyl-CCNU-magnetic emulsion (no magnet)	8	19.9 ± 2.2	113
Methyl-CCNU-magnetic emulsion (with magnet)	8	22.5 ± 1.2	128

[a] The ratio of the mean survival time of the treated group (T) to that of the control group (C).
[b] Control group had no chemotherapy.
[c] Intraperitoneal injection.

2. Antitumor Effect of ACNU Entrapped in Magnetic Emulsions Against AH 7974 Lung Metastasis

The effect of ACNU when administered as a magnetic emulsion on the AH 7974 lung metastasis is shown in Table 4. The efficacy of ACNU was measured with rats injected intravenously with a single dose of 10 mg/kg ACNU. The survival time in rats for free ACNU group was not markedly different from that of control rats. In contrast, after magnetic emulsions containing ACNU were injected into rats applied with an electromagnet, the mean survival was 23.1 days and T/C was 131%, the value being higher than that in the control group ($p<0.01$) or free ACNU group ($p<0.01$). However, the survival time and T/C in rats receiving magnetic emulsions containing ACNU without application of magnetic field was 22.1 days and 126%, respectively. There was no significant difference in survival time between both groups of rats given magnetic emulsions containing ACNU with and without the application of electromagnet.

D. Comment

Magnetic emulsions prepared with oleic acid- or ethyl oleate-based magnetic fluids were stable and showed the potent in vitro and in vivo magnetic responsiveness. Oil in water magnetic emulsions are suitable to entrap lipophilic compounds while magnetic albumin microspheres are to water soluble drugs. Other types of emulsions such as water in oil and multiples (o/w/o, w/o/w) may be used to overcome the specificity for entrapping agents. In this experiment, o/w magnetic emulsions are shown to be especially useful for protection and stabilization of unstable agents, (i.e., lipophilic nitrosoureas) against the biological fluids. In addition, magnetic emulsions may be useful as an adjuvant chemotherapy during surgical operations and as an assistant therapy before and after operations. Also magnetic emulsions are a good candidate as a chemoembolizing agent of the small arteries into the tumor regions.[58]

IV. CONCLUSION

From the results and discussion presented, it was suggested that magnetic albumin microspheres and magnetic (o/w) emulsions were useful for the targeting of entrapped drugs (antitumor agents) into a predetermined site of body. Since it is a reasonable interpretation that magnets attract one half of ferrites and draw another half apart, it is very difficult to concentrate all of the ferrites on magnets. Unless a very small and

Table 4
EFFECT OF VARIOUS FORMULATIONS OF ACNU ON SURVIVAL TIME OF RATS BORN WITH AH 7974 LUNG CARCINOMA

Formulation	Number of rats	Mean survival time ± SE (days)	T/C%[a]
Control	16	17.6 ± 2.6	100
ACNU-aqueous solution	8	18.0 ± 1.2	102
Magnetic emulsion[b] (with magnet)	5	17.2 ± 1.3	98
ACNU-magnetic emulsion (no magnet)	8	22.1 ± 1.4	126
ACNU-magnetic emulsion (with magnet)	10	23.1 ± 0.3	131

[a] See Table 3.
[b] This formulation does not contain chemotherapeutic agents.

powerful magnet is available, moreover, it is not easy to target magnetic drug carriers on a predetermined site. In general, a small magnet has poor magnetic power. Therefore, to find a small and powerful magnet becomes an important factor for targeting drug delivery by means of magnet and magnetic drug carriers. This problem may be solved if we push electronics industries to make a powerful magnet by producing data on the usefulness of targeting antitumor agents using magnetic drug carriers. In our laboratory, we are estimating more utility of nonmagnetic albumin microspheres after intra-arterial administration in rats and rabbits, with expectation of an embolization with microspheres in a predetermined site of a small blood vessel and a sustained release of antitumor agents from microspheres. Superselective catheterization and magnet/magnetic drug carriers may help an attempt for enhancement of localization of carriers and entrapped agents. In any case, the experiments with magnetically responsive drug carriers which we have done are not enough and clinical applications are dependent upon further experiments.

REFERENCES

1. Tyrell D. A., Health, T. D., Colley, C.M., and Ryman, B. E., New aspects of liposomes, *Biochim. Biophys. Acta,* 457, 259, 1976.
2. Juliano, R. L. and Stamp, D., Pharmacokinetics of liposome entrapped anti-tumor drugs; studies with vinblastine, actinomycin D, cytosine arabinoside and daunomycin, *Biochem. Pharmacol.,* 27, 21, 1978.
3. Kimelberg, H. K., Differential distribution of liposome - entrapped [^{3}H] methotrexate and labelled lipids after intravenous injection in a primate, *Biochim. Biophys. Acta,* 448, 531, 1976.
4. Tanaka, T., Taneda, K., Kobayashi, H., Okumura, K., Muranishi, S., and Sezaki, H., Application of liposomes to the pharmaceutical modification of the distribution characteristics of drug in the rat, *Chem. Pharm. Bull.,* 23, 3069, 1975.
5. Gregoriadis, G. and Ryman, B. E., Fate of protein-containing liposomes injected into rats. An approach to the treatment of storage diseases, *Eur. J. Biochem.,* 24, 485, 1972.
6. Meyhew, E., Papahadjopoulos, D., Rustum, Y. M., and de Duve, C., Inhibition of tumor cell growth *in vitro* and *in vivo* by 1-β-D-arabinofuranosyl cytosine entrapped within phospholipid vehicles, *Cancer Res.,* 36, 4406, 1976.
7. Trouet, A., Compeneere, D. D., and de Duve, C., Chemotherapy through lysosomes with a DNA daunorubicin complex, *Nature (London),* 239, 110, 1972.

8. Trouet, A., Campeneere, D. D., Smedt-Meleugreaux, M. S., and Attassi, G., Experimental leukemia chemotherapy with a "lysosomotropic" adriamycin-DNA complex, *Eur. J. Cancer,* 10, 405, 1974.
9. Sugibayashi, K., Morimoto, Y., Nadai, T., and Kato, Y., Drug carrier property of albumin microspheres in chemotherapy. I. Tissue distribution of microsphere-entrapped 5-fluorouracil in mice, *Chem. Pharm. Bull.,* 25, 3433, 1977.
10. Sugibayashi, K., Morimoto, Y., Nadai, T., Kato, Y., Hasegawa, A., and Arita, T., Drug-carrier property of albumin microspheres in chemotherapy. II. Preparation and tissue distribution in mice of microsphere-entrapped 5-fluorouracil, *Chem. Pharmacobio. Bull.,* 27, 204, 1979.
11. Sugibayashi, K., Akimoto, M., Morimoto, Y., Nadai, T., and Kato, Y., Drug carrier property of albumin microspheres in chemotherapy. III. Effect of microsphere- entrapped 5-fluorouracil on Ehrlich ascites carcinoma in mice, *J. Pharmacobio. Dyn.,* 2, 350, 1979.
12. Morimoto, Y., Akimoto, M., Sugibayashi, K., Nadai, T., and Kato, Y., Drug-carrier property of albumin microspheres in chemotherapy. IV. Anti-tumor effect of one or multiple shot of microsphere-entrapped 5-fluorouracil on Ehrlich ascites or sold tumor in mice. *Chem. Pharm. Bull.,* 28, 3087, 1980.
13. Morimoto, Y., Sugibayashi, K., and Kato, Y., Drug-carrier property of albumin microspheres in chemotherapy. V. Antitumor effect of microsphere-entrapped adriamycin on liver metastasis of AH 7974 cells in rats, *Chem. Pharm. Bull.,* 29, 1433, 1981.
14. Turner, R. D., Rand, R. W., Bentson, J. R., and Mosso, J. A., Ferromagnetic silicone necrosis of hypernephromas by selective vascular occlusion to the tumor. A new technique, *J. Urol.,* 113, 455, 1975.
15. Kato, T., Enhancement of antitumour effects by magnetic control of microcapsulated anticancer drug, *Gan to Kagakuryoho,* 8, 698, 1980.
16. Widder, K. J., Senyei, A. J., and Scarpelli, D. G., Magnetic microspheres. A model system for site specific drug delivery *in vivo, Proc. Soc. Exp. Biol. Med.,* 158, 141, 1978.
17. Mosbach, K. and Schroder, U., Preparation and application of magnetic polymers for targeting of drugs, *FEBS Lett.,* 102, 112, 1979.
18. Ibrahim, A., Couvreur, P., Roland, M., and Speiser, P., New magnetic drug carrier, *J. Pharm. Pharmacol.,* 35, 59, 1983.
19. Morimoto, Y., Sugibayashi, K., Okumura, M., and Kato, Y., Biomedical applications of magnetic fluids. I. Magnetic guidance of ferro-colloid-entrapped albumin microsphere for site specific drug delivery *in vivo, J. Pharmacobio. Dyn.,* 4, 624, 1981.
20. Morimoto, Y., Okumura, M., Sugibayashi, K., and Kato, Y., Biomedical applications of magnetic fluids. II. Preparation and magnetic albumin microsphere for site specific drug delivery *in vivo, J. Pharmacobio. Dyn.,* 4, 624, 1981.
21. Sugibayashi, K., Okumura, M., and Morimoto, Y., Biomedical applications of magnetic fluids. III. Antitumor effect of magnetic albumin microsphere-entrapped adriamycin on lung metastasis of AH 7974 in rats, *Biomaterials,* 3, 181, 1982.
22. Akimoto, M. and Morimoto, Y., The development of a magnetic emulsion as drug carrier, *J. Pharmacobio. Dyn.,* 5, s-15, 1982.
23. Morimoto, Y., Akimoto, M., and Yotsumoto, Y., Dispersion state of protein-stabilized magnetic emulsions, *Chem. Pharm. Bull.,* 30, 3024, 1982.
24. Akimoto, M. and Morimoto, Y., Use of magnetic emulsion as a novel drug carrier for chemotherapeutic agents, *Biomaterials,* 4, 49, 1983.
25. Morimoto, Y., Sugibayashi, K., and Akimoto, M., Magnetic guidance of ferro-colloid-entrapped emulsion for site - specific drug delivery, *Chem. Pharm. Bull.,* 31, 279, 1983.
26. Akimoto, M., Sugibayashi, K., and Morimoto, Y., Application of magnetic emulsions for sustained release and targeting of drugs in cancer chemotherapy, *J. Controlled Resease,* 1, 205, 1985.
27. Shimoiizaka, J., Nakatsuka, K., Chubachi, R., and Sato, Y., Stabilization of aqueous magnetic suspension. Preparation of water-based magnetic fluid, *Nippon Kagaku Kaishi,* 1976, 6, 1976.
28. Scheffel, U., Rhodes, B. A., Natarajan, T. K., and Wagner, H. N., Jr., Albumin microspheres for study of reticuloendothelial system, *J. Nucl. Med.,* 13, 498, 1972.
29. Fujimoto, S., Preparation and application of albumin microspheres containing antitumor agents as a drug carrier for chemembolization, presented at the 16th Symp. on Drug Metabolism and Action, Gifu, Japan, November 1984.
30. Kramer, P. A., Albumin microspheres as vehicles for achieving specificity in drug delivery, *J. Pharm. Sci.,* 63, 1646, 1974.
31. Lee, T. K., Sokoloski, T. D., and Royer, G. P., Serum albumin beads. An injectable, biodegradable system for the sustained release of drugs, *Science,* 213, 233, 1981.
32. Kubioatowicz, D. O., Ithakissios, D. S., Wicks, J. H., and Heerwald, P. E., Semi-automated T_3 uptake test that uses magnetic albumin microspheres, *J. Nucl. Med.,* 19, 854, 1978.
33. Widder, K., Flouret, G., and Senyei, A., Magnetic microspheres. Synthesis of a novel parenteral drug carrier, *J. Pharm. Sci.,* 68, 79, 1979.

34. Knop, G., Microspheres of human serum albumin, East German Patent, 115, 037, 1975 (C.A., 86, 145921m, 1977).
35. Martodam, R. R., Twumasi, D. Y., Liener, I. E., Powers, J. C., Nishino, N., and Kurejcarek, G., Albumin microspheres as carrier of an inhibitor of leukocyte elastase. Potential therapeutic agent for emphysema, *Proc. Natl. Acad. Sci. U.S.A.*, 76, 2128, 1979.
36. Rhodes, B. A., Stern, H. S., Buchanan, J. W., Zolle, I., and Wagner, H. N., Jr., Lung scanning with ^{99m}Tc-microspheres, *Radiology*, 99, 613, 1971.
37. Rhodes, B. A. and Bolles, T. F., Albumin microspheres. Current methods of preparation and use, in *Radiopharmaceutics*, Subramarian, C., Cooper, J. F., Rhodes, B. A., and Sodd, V. J., Eds., Society of Nuclear Medicine, New York, 1975, 282.
38. Bates, D. V., Macklem, P. T., and Christie, R. V., in *Respiratory Function and Disease*, W. B. Saunders, Philadelphia, 1971, 367.
39. Jones, J. G. and Wagner, C. G., Chronic exposure to iron oxide, chromium oxide and nickel oxide fumes of metal dressers in a steelworks, *Br. J. Ind. Med.*, 29, 169, 1972.
40. Cohen, D., Ferromagnetic contamination in the lungs and other organs of the human body, *Science*, 180, 745, 1973.
41. Tanri, T. and Watanabe, T., Private communication, 1976.
42. Tobe, M., Kobayashi, K., Suzuki, S., and Ikeguchi, Y., private communication, 1972.
43. Senyei, A. E., Reich, S. D., Gonczy, C., and Widder, K. J., *In vivo* kinetics of magnetically targeted low-dose doxorubicin, *J. Pharm. Sci.*, 70, 389, 1981.
44. Mimran, A. and Casellas, D., Microsphere size and determination of intrarenal blood flow distribution in the rats, *Pfluegers Arch.*, 382, 233, 1979.
45. Kanke, M., Simmons, G. H., Weiss, D. L., Bivins, B. A., and Deluca, P. P., Clearance of ^{141}Ce-labeled microspheres from blood and distribution in specific organs following intravenous and intraarterial administration in beagle dogs, *J. Pharm. Sci.*, 69, 755, 1980.
46. Sjoholm, I. and Edman, P., Acrylic microspheres *in vivo*. I. Distribution and elimination of polyacrylamide microparticles after intravenous and intraperitoneal injection in mouse and rat, *J. Pharmacol. Exp. Ther.*, 211, 656, 1979.
47. Widder, K. J., Senyei, A. E., and Ranney, D. F., Magnetically responsive microspheres and other carriers for the biophysical targeting of antitumor agents, in *Advances in Pharmacology and Chemotherapy*, Vol. 16, Garottini, S., Goldin, A., Hawking, F., Koplin, I. J., and Schnitzer, R. J., Eds., Academic Press, New York, 1979, 213.
48. Gregoriadis, G., Targeting of drugs, *Nature (London)*, 265, 407, 1977.
49. Gregoriadis, G., Ed., *Drug Carriers in Biology and Medicine*, Academic Press, New York, 1979.
50. Gregoriadis, G., Tailoring liposome structure, *Nature (London)*, 283, 814, 1980.
51. Odashima, S., Establishment of ascites hepatomas in the rat, 1951—1962, in Ascites Tumors - Yoshida Sarcoma and Ascites Hepatama(s), Yoshida, T., Ed., *Natl. Cancer Inst. Monogr.*, 16, 51, 1964.
52. Ichimura, H., Experimental studies on a screening system of anticancer drugs employing the rat ascites hepatoma spectrum, *Gan to Kagakuryoho*, 2, 605, 1975.
53. Gregoriadis, G., Swain, C. P., Wills, E. J., and Tavill, A. S., Drug carrier potential of liposomes in cancer chemotherapy, *Lancet*, 1, 1313, 1974.
54. Hashida, H., Muranishi, S., Sezaki, H., Tanigawa, N., Satomura, K., and Hikasa, Y., Increased lymphatic delivery of bleomycin by microsphere in oil emulsion and its effect on lymph node metastasis, *Int. J. Pharm.*, 2, 245, 1979.
55. Morimoto, Y., Akimoto, M., and Sugibayashi, K., Stability of nitrosourea antitumor agents in aqueous solution and plasma, in preparation.
56. Loo, T. L. and Dion, R. L., Colorimetric method for the determination of 1,3-bis (2-chloroethyl)-1-nitrosourea, *J. Pharm. Sci.*, 54, 809, 1965.
57. Nakamura, K., Asami, M., Kawada, K., and Sasahara, K., Quantitative determination of ACNU (3-[(4-amino-2-methyl-5-pyrimidyl)methyl]-1-(2-chloroethyl)-1-nitrosourea hydro-chloride), a new water-soluble anti-tumor nitrosourea, in biological fluids and tissues of patients by high - performance liquid chromatography. I. Analytical method and pharmacokinetics, *Ann. Rep. Sankyo Res. Lab.*, 29, 66, 1977.
58. Fujimoto, S., Endoh, F., Okui, K., Morimoto, Y., Sugibayashi, K., Miyagawa, A., and Suzuki, H., Continued *in vitro* and *in vivo* release of an antitumor drug for albumin microspheres, *Experimentia*, 39, 913, 1983.

Chapter 4

OCULAR THERAPY WITH NANOPARTICLES

Robert Gurny

TABLE OF CONTENTS

I. INTRODUCTION

Frequent dosage is probably the major reason for noncompliance with prescribed schedules of administration. Moreover, especially in the case of ocular therapy, the precorneal disposition factors for all ocular drugs increase the difficulty of preparing an acceptable controlled-release formulation. Unlike most systemic therapies, the major portion of a topically instilled drug leaves the absorption site (cornea) unabsorbed. This occurs because of tear dilution and a washout phenomenon.[1,2] While drugs in saline solutions are commonly used in opthalmic practice, solutions with viscosities of 1 to 50 cps are used in order to improve comfort of lubrication and also to achieve a slight prolongation in action. In recent years, different types of ocular therapeutic systems have been extensively tested and improved:

1. Reservoir type systems
2. Soluble or insoluble inserts
3. Emulsions
4. Liposomes
5. Implantable pumps
6. Latex formulations

Among these new types of delivery systems are submicroscopic dispersions for ophthalmic use which have been tested in animals and man in recent years.[3-7] The authors call these new therapeutic systems latices, pseudolatices, or emulsions and have reported promising results in ocular therapy.

In pharmaceutics, most of the researchers make no distinction between latices and pseudolatices, since the final products have the same characteristics. However, in the first case, the dispersion is prepared by the well-known technique of emulsion polymerization, while the second one is a typical mechanical dispersion of already existing polymer particles.[8]

II. CLASSIFICATION OF LATEX SYSTEMS

Vanderhoff et al.[9] developed one of the basic technologies of preparing these colloidal dispersions from already formed polymers. The impossibility of obtaining latex systems by emulsion polymerization from a large number of compounds, e.g., epoxy resins, polyurethanes, polyesters, cellulosic derivatives such as ethylcellulose, and elastomers such as cis-polyisoprene, led to this new technique. This method of preparing aqueous dispersions has already found an interesting application in the field of aqueous coating.[10]

Up to now, three distinct methods may be mentioned:[11] (1) selfemulsification; (2) phase inversion; (3) solution emulsification. Recent research[12,13] suggest the possibility of using mixed emulsifying systems as potent stabilizers, e.g., laurylsulfate-cetylalcohol, hexadecyltrimethyl ammonium bromide-cetylalcohol, or steric stabilizers, e.g., polyoxyethylene.

Figure 1 shows the main difference between the two basic methods for preparing latex systems. The one on the left is a typical emulsion polymerization, the one on the right is a dispersion of an already formed polymer in water.

In general, the three main categories of latex products may be distinguished according to their origins:[11]

1. Natural latices — which occur as the metabolic products of various plants and trees.

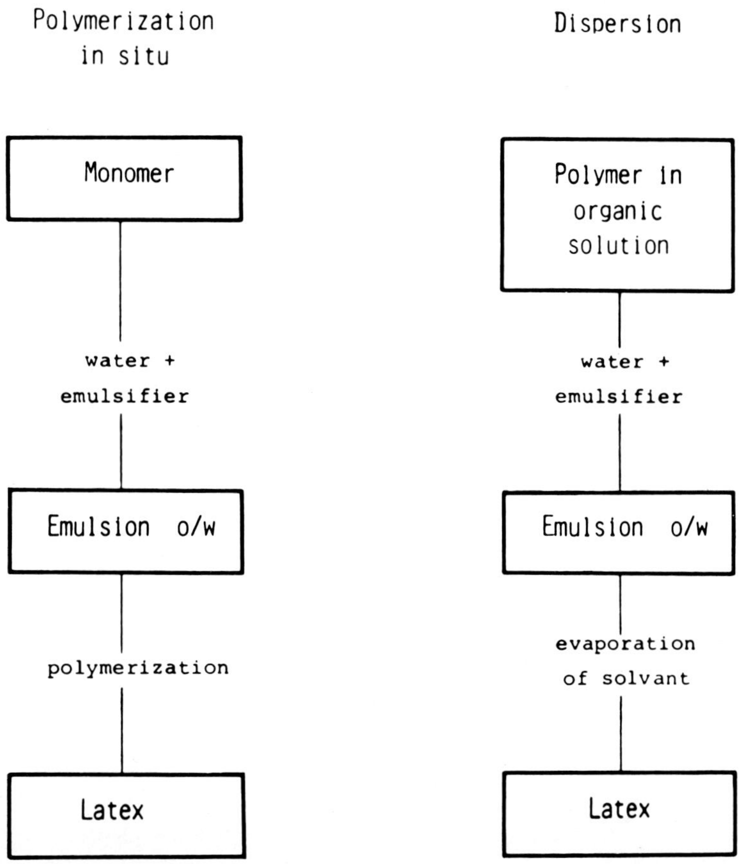

FIGURE 1. Methods of latex preparation. (From Gurny, R., in *Topics in Pharmaceutical Sciences,* Breimer D. D. and Speiser P., Eds., Elsevier, Amsterdam, 1983, 277. With permission.)

2. Synthetic latices — prepared directly from their corresponding monomers by emulsion polymerization.
3. Artificial latices (pseudolatices) — prepared by dispersion of bulk polymers already formed in aqueous media.

In order to use these submicroscopic dispersions as drug delivery systems in the eye, two possible means for incorporation of the active ingredient are feasible: (1) incorporation during manufacture of the latex or (2) adsorption on the surface once the latex is prepared.

III. THE USE OF NANOPARTICLES IN OCULAR THERAPY

A. Polymer Emulsions

Ticho et al.[3] worked on an aqueous polymer emulsion to be used as pilocarpine-releasing eye drops for the treatment of glaucoma. The active ingredient is chemically bound to the polymer.

Such a system, containing 3.4% pilocarpine as a polymeric salt has been evaluated in numerous studies.[3,5,6,13-17]

The in vitro release patterns of such systems[6] were studied by determining the amount of pilocarpine liberated from a dialysis bag containing the latex (pilocarpine

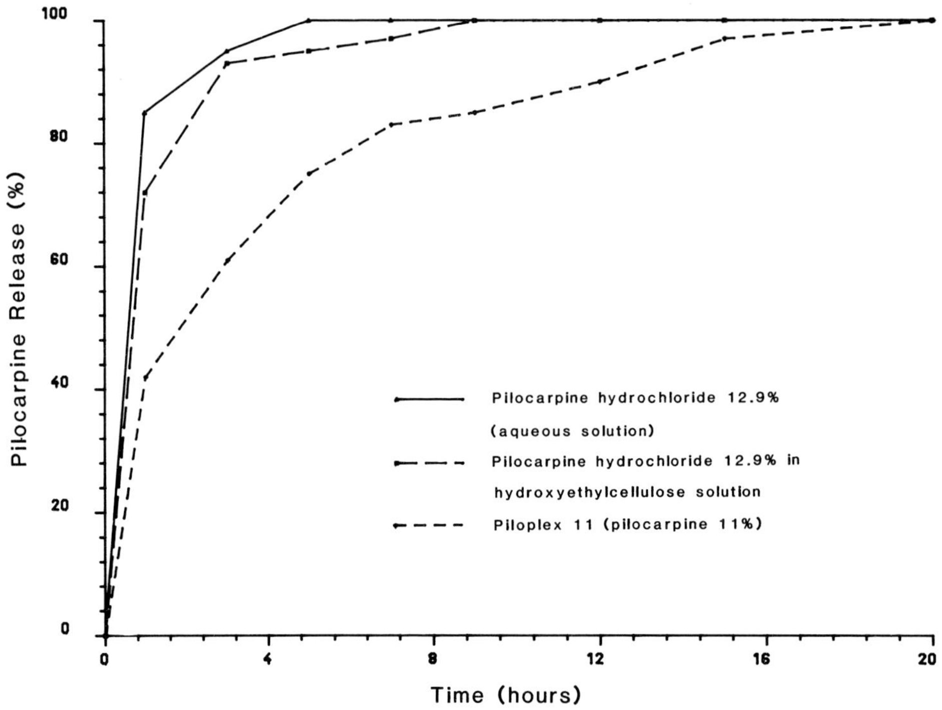

FIGURE 2. In vitro release of pilocarpine from a polymeric salt (Piloplex®) and pilocarpine hydrochloride solution, from a dialysis bag. The medium was an agitated isotonic saline solution (pH 7 .1), maintained at 37°C. The active ingredient was determined spectrophotometrically.[6]

base or aqueous solutions of pilocarpine hydrochloride). The release medium was an agitated isotonic saline solution (pH 7.1) maintained at 37°C. The results, summarized in Figure 2, indicate that the release time of 80% of the active ingredient from the polymeric salt is 6 hr against 1 hr for the corresponding solutions.

Figure 3 shows the average diurnal intraocular pressure (IOP) of nine eyes on the third day of pilocarpine treatment with the emulsion in comparison with a 4% solution of pilocarpine. The latex system (emulsion) had a lower level of pilocarpine (3.4% w/v) and showed less fluctuation than the corresponding solution (4% w/v). Data concerning the volunteer subjects treated show a reduction of 5.25 mmHg in the average diurnal value.

Throughout a 1 year study only 1 patient out of 30 with open angle glaucoma complained of a local sensitivity reaction. The very promising results collected over a period of 8 to 12 months are given in Figure 4. Out of 15 patients studied, 87% had an IOP value of less than 24 mmHg throughout all measurement and no patient showed progression in field loss.

In another large study involving almost 50 patients, the comparative evaluation of the polymeric system and a pilocarpine solution against no treatment is shown (Figure 5). The partial results of the extensive study show that over 80% of the eyes treated with a long acting system (Piloplex® 11) had a lower mean IOP value than the control group treated with an aqueous solution of pilocarpine (2% solution). Again, the diurnal IOP value of the group treated with the polymeric salt was significantly lower ($p<0.005$) than the one treated with the solution by 2.32 mmHg. 67.6% of the eyes treated with the new therapeutic system were under control (IOP values $\leq$20 mmHg), whereas only 45.2% were under control with the ordinary pilocarpine treatment.

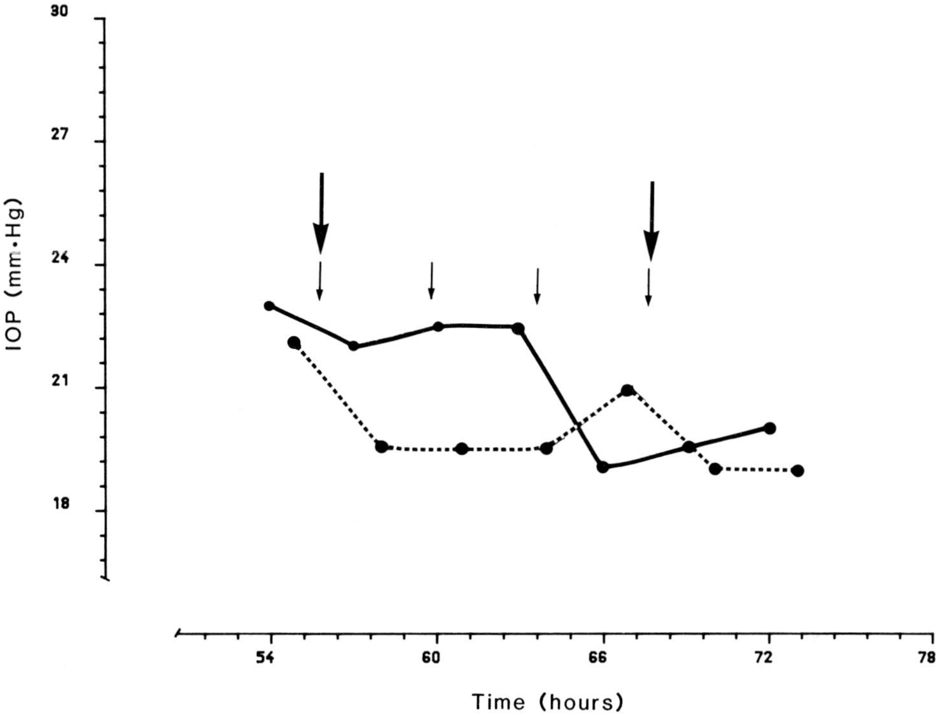

FIGURE 3. Average diurnal IOP curves of 9 eyes on the 3rd day of pilocarpine hydrochloride 4% treatment (——) and on the 3rd day of polymer emulsion (Piloplex® 3.4) treatment (- - - - - -), time of pilocarpine solution application (↓), time of emulsion application (⬇).[3]

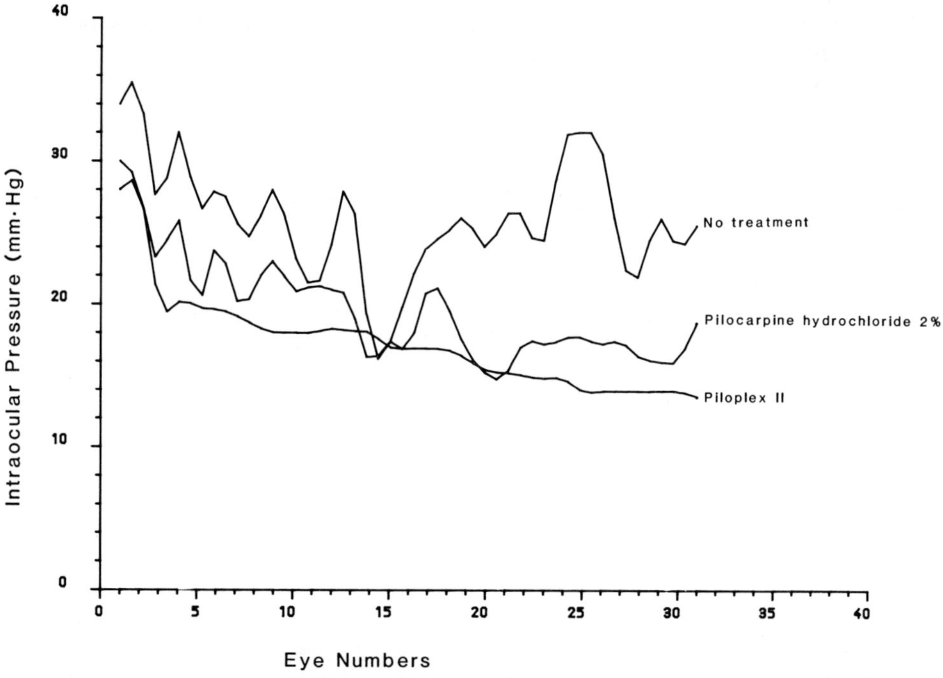

FIGURE 4. Average morning IOP curve of the 30 eyes studied using pilocarpine hydrochloride 4 times daily followed by Piloplex® twice daily treatment.[6]

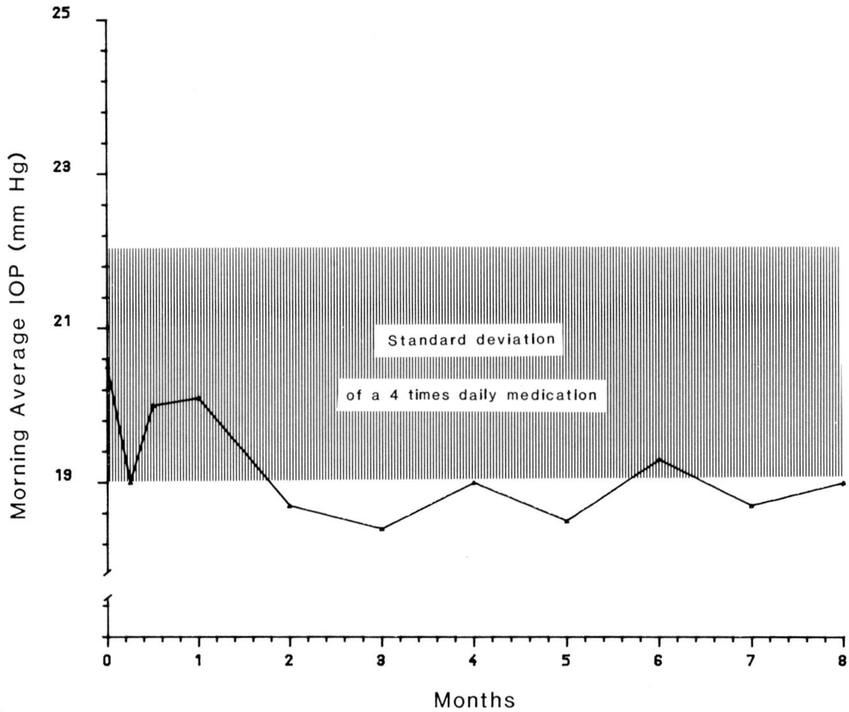

FIGURE 5. Average diurnal IOP values of 31 eyes in 3 different periods. The average values were calculated from the diurnal IOP values on the 2nd day of no treatment period and the 3rd day each of the pilocarpine hydrochloride 2% treatment and the polymeric salt (Piloplex® 11) treatment.[3]

There have been no objective signs of damage and specifically, neither corneal abrasion nor intraocular inflammatory signs were present.

B. Latices

The first report on a latex system for ocular delivery with variable viscosity in function of the pH was in 1980.[4] Since then, various papers[7,18-20] have been published on the preparation of these formulations and their use for the controlled release of drugs in the eye.

These drug delivery systems are based on cellulose acetate hydrogen phthalate (CAP), used widely in the pharmaceutical industry. The cellulose derivative in the latex starts to dissolve at a pH of about 5.0.

The CAP latices containing the active compound (pilocarpine) adsorbed partially onto the surface of the colloidal polymeric beads of the dispersions have a pH of 4.5 and show Brookfield viscosities between 50 and 200 cps. The particular behavior of these latices is to coagulate as soon as they are applied in the cul-de-sac, since the lacrimal fluid has a pH of 7.2. The gel thus formed from the latex has a Brookfield viscosity of several tenfolds above the initial viscosity. The gelified form cannot be washed out by the lacrimal fluid. Figure 6 gives a close view of the coagulation process in pictures taken with a scanning electron microscope. Within a few seconds after contact of the preparation with the tear fluid which has a pH 2.8 units above the one of the preparation, the surface of the polymeric beads starts to dissolve.

The long-acting latices, once coagulated, have no effect on the visual acuity. The relative miotic response over time of these new therapeutic systems in comparison to solutions is given in Figure 8.

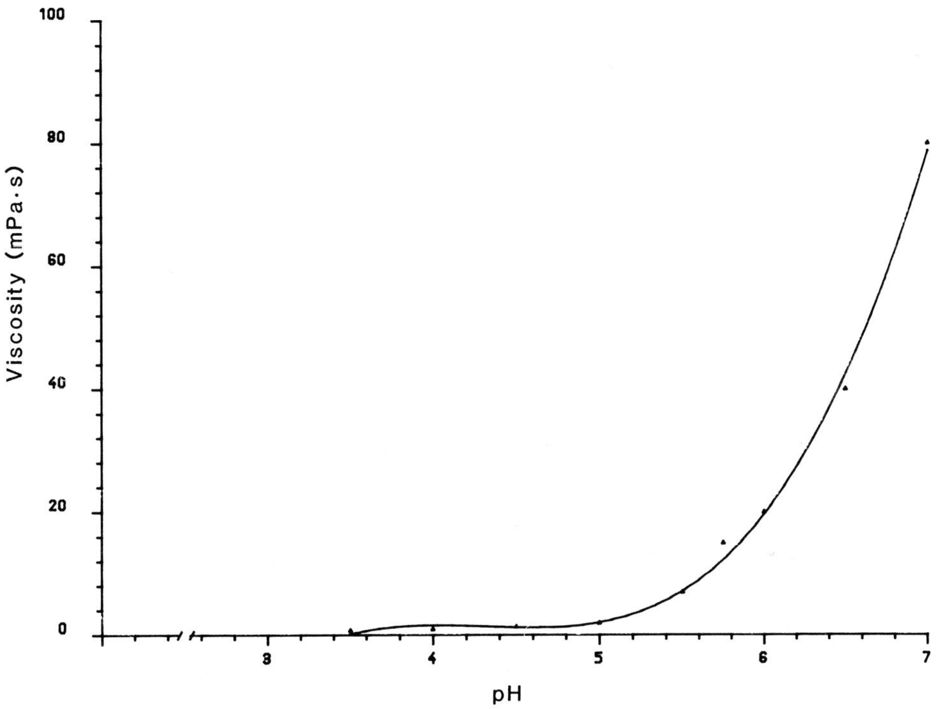

FIGURE 6. Plot of the viscosity behavior of a diluted CAP latex (4% w/v) vs. pH, showing rapid increase in viscosity due to gel formation.

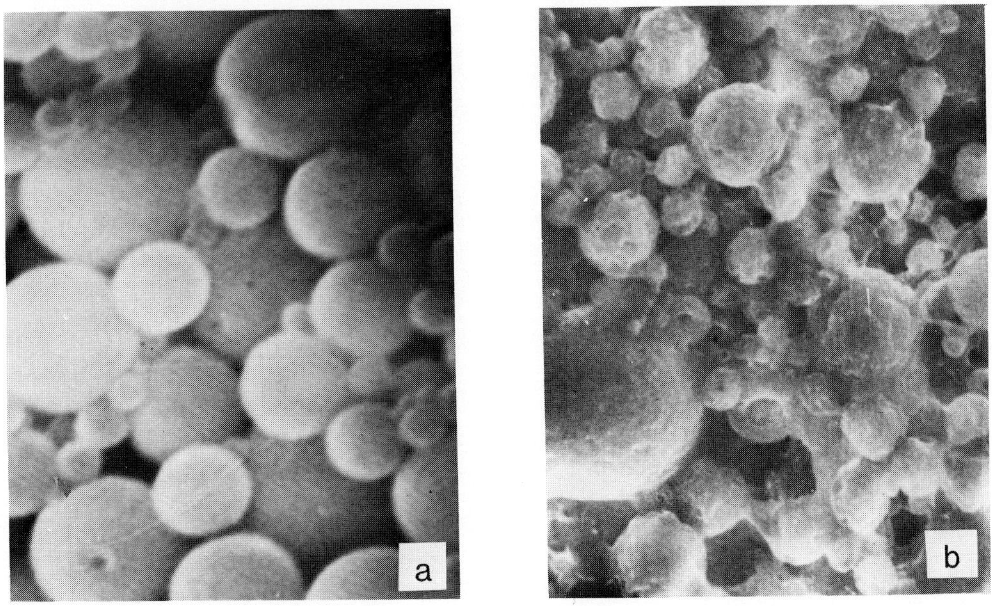

FIGURE 7. Transformation of a latex into a gel due to pH change. (a) Stable latex, (b) beginning of coalescence and gelification. (From Gurny, R., in *Topics in Pharmaceutical Sciences,* Breimer, D. D. and Speiser, P., Eds., Elsevier, Amsterdam, 1983, 277. With permission.)

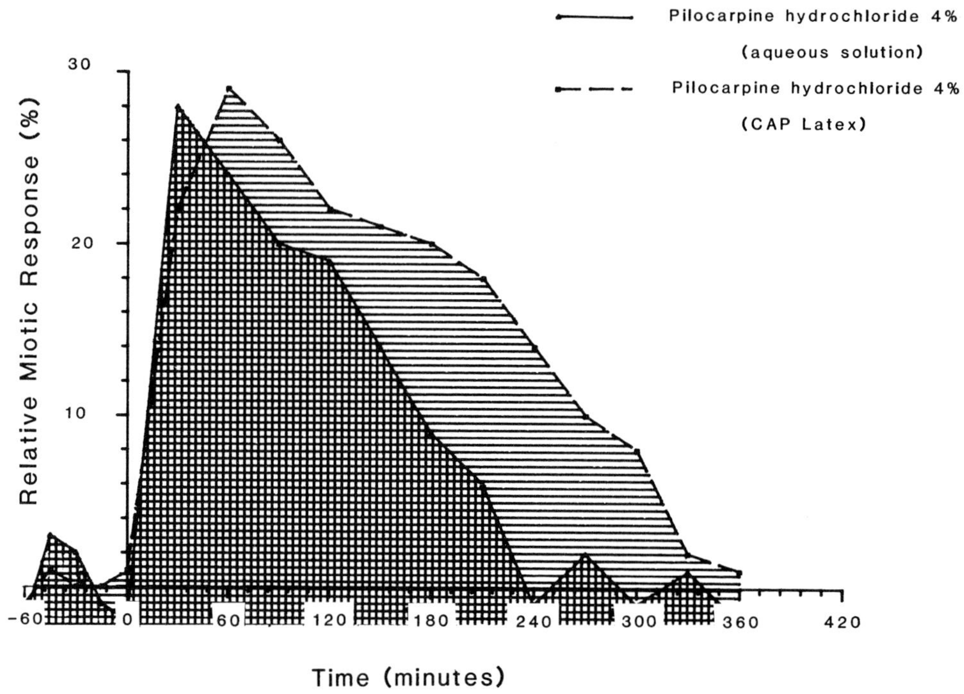

FIGURE 8. Comparison of the miotic response of two dosage forms. Solution of pilocarpine · HCl (4%) (——); latex CAP with pilocarpine · HCl (4%) (- - -).

Table 1
AVERAGE PHARMACOKINETIC VALUES OF MIOSIS IN SIX RABBITS AFTER INSTILLATION OF A SOLUTION AND A LATEX

Parameter	Latex system	Solution
Active ingredient % (w/v)	4	4
Polymer content % (w/v)	30	—
AUC (% · min)	5214 (1040)	3396 (835)
RI_{max} (%)	27.4 (1.5)	27.3 (1.5)
t_{max}	60 (—)	30 (—)
$\Delta_{1/2}$ (min)	198 (12)	132 (18)

Note: Numbers in parentheses are standard deviation.

The delivery device contains 30% w/v polymer and the active ingredient is at the level of 4% w/v. Table 1 gives the area under the curve (AUC), time to peak (t_{max}), the maximum relative response intensity of miosis (RI_{max}), and the duration at half the maximum response $\Delta_{1/2}$).

The comparative values for other drug concentrations are plotted in Figure 9 for ordinary solutions as well as colloidal dispersions.

The correlation between AUC and the drug concentration shows clearly the interesting features of these new systems. Furthermore, the interesting comparison of the values $\Delta_{1/2}$ (duration of response at half the maximum level) is given in Figure 10.

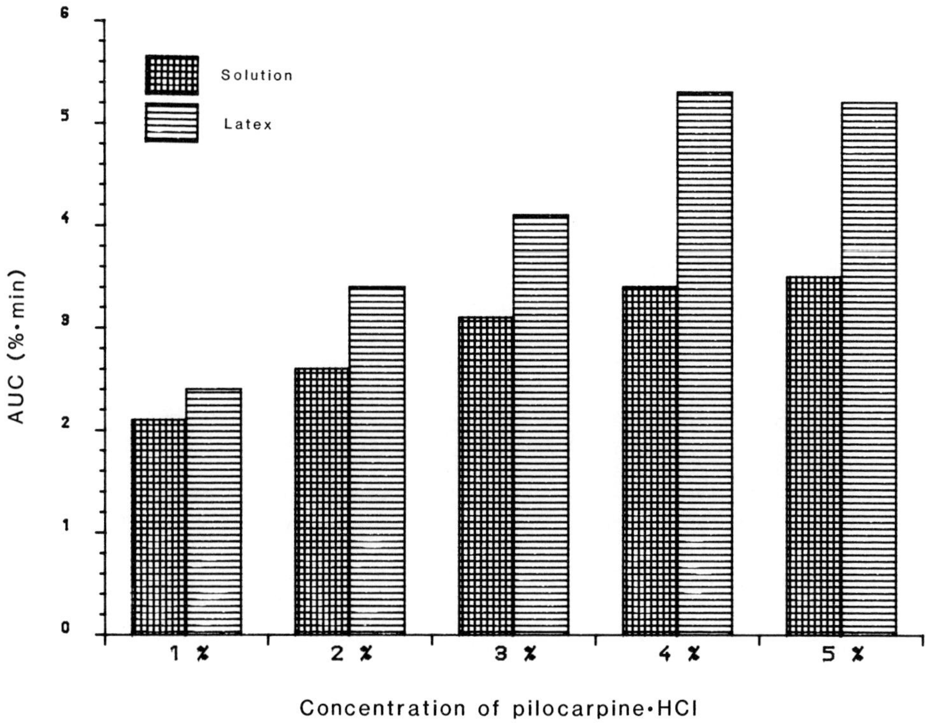

FIGURE 9. Comparison of the AUC values between ordinary solutions of pilocarpine HCl and latices containing between 1 and 5% w/v of active material.

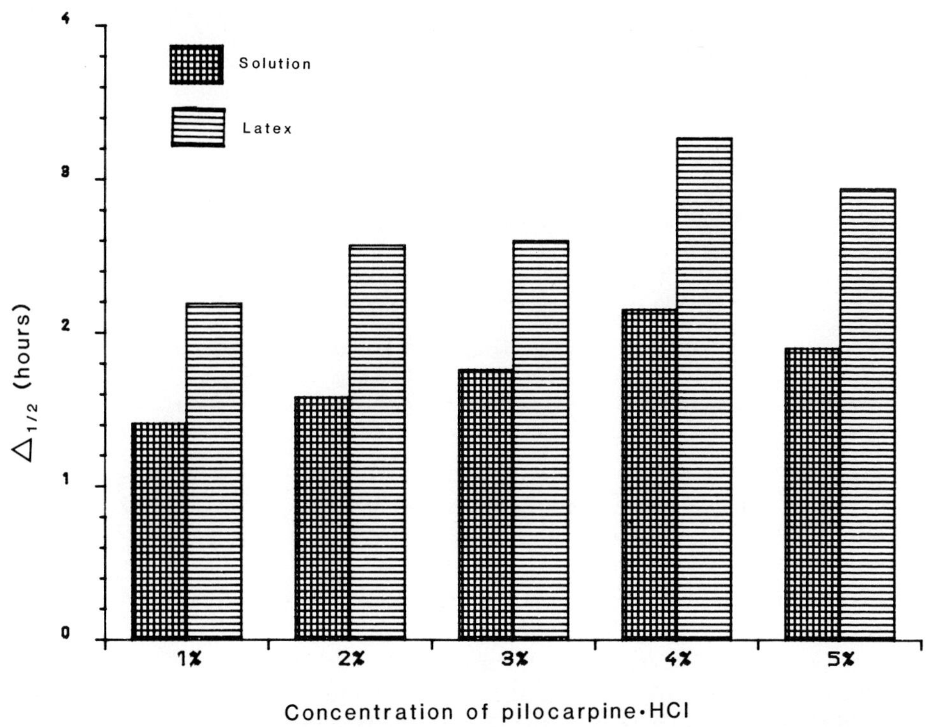

FIGURE 10. Comparison of the $\Delta_{1/2}$ values between solutions and latices containing 1 to 5% w/v of pilocarpine HCl.

REFERENCES

1. Chrai, S. S., Makoid, M. C., Eriksen, S. P., and Robinson, J. R., Drop size and initial dosing frequency problems of topically applied ophthalmic drugs, *J. Pharm. Sci.*, 63, 333, 1974.
2. Chrai, S. S., Patton, T. F., Mehta, A., and Robinson, J. R., Lacrimal and instilled fluid dynamics in rabbit eyes, *J. Pharm. Sci.*, 62, 1112, 1973.
3. Ticho, U., Blumenthal, M., Zonis, S., Gal, A., Blank, I., and Mazor, Z. W., A clinical trial with Piloplex — a new long — acting pilocarpine compound: preliminary report, *Ann. Ophthalmol.*, 11, 555, 1979.
4. Gurny, R. and Taylor, D., Development and evaluation of a prolonged acting drug delivery system for the treatment of glaucoma, in *Proc. Int. Symp. of the British Pharmaceutical Technology Conference*, Rubinstein, M. H., Ed., Solid Dosage Research Unit, Liverpool, England, 1980.
5. Mazor, Z., Ticho, U., Rehany, U., and Rose, L., Piloplex, a new long-acting pilocarpine polymer salt. B: comparative study of pilocarpine and Piloplex eye drops, *Br. J. Ophthalmol.*, 63, 48, 1979.
6. Ticho, U., Blumenthal, M., Zonis, S., Gal, A., Blank, I., and Mazor, Z. W., Piloplex, a new long-acting pilocarpine polymer salt. A: long-term study, *Br. J. Opthalmol.*, 63, 45, 1979.
7. Gurny, R., Preliminary study of prolonged acting drug delivery system for the treatment of glaucoma, *Pharm. Acta Helv.*, 56, 130, 1981.
8. Bindschaedler, C., Gurny, R., and Doelker, E., Notions théoriques sur la formation des films obtenus à partir de microdispersions aqueuses et application à l'enrobage, *Labo-Pharma — Probl. Tech.*, 31, 389, 1983.
9. Vanderhoff, J. W., El-Asser, M. S., and Ungelstad J., U.S. Patent, 4,177,177.
10. Gurny, R., The use of pseudolatices for aqueous filmcoating, in *Proc. Int. Symp. of Pharmaceutical Technology and Product Manufacture*, Goldberg, A. S. Ed., Powder Advisory Centre, London, England, 1979.
11. El-Asser, M. S., in *Int. Symp. of Advances in Emulsion Polymerization and Latex Technology*, Vol. 2, El-Asser, M. S., Ed., Lehigh University, Bethlehem, Pa., 1979, chap. 19.
12. Gurny, R., Peppas, N. A., Harrington, D. D., and Banker, G. S., Development of biodegradable and injectable latices for controlled release of potent drugs, *Drug. Dev. Ind. Pharm.*, 7, 1, 1981.
13. Mazor, Z., Kazan, R., Kain, N., Ladkani, D., Ross, M., and Weiner, B., Glaubid(R) (Piloplex 3,4) — A long-acting, anti-glaucoma medication, presented at Int. Symp. on Glaucoma, Jerusalem, August 15, 1983.
14. Bonomi, L., Perfetti, S., Bellucci, R., and Massa, F., Studio sperimentale dell'attivita di un polimero di pilocarpina a lunga durata d'azione, *Boll. Ocul.*, 60, 909, 1981.
15. Robinson, J. R. and Li, V. H. K., Ocular disposition and bioavailability from Piloplex and other drug delivery systems, present at Int. Symp. on Glaucoma, Jerusalem, August 15, 1983.
16. Duzman, E., Irvine, U. C., and Irvine, C. A., A one-month crossover trial comparing the safety and intraocular pressure control in patients treated with Piloplex 3.4 applied twice daily with those treated with pilocarpine 2% applied four times daily, presented at Int. Symp. on Glaucoma, Jerusalem, August 15, 1983.
17. Ticho, U., Brandeis, E., Lazer, M., Lifshitz, Y., Geyer, O., Ben-Sira, Y., Lusky, M., Cohen, S., Shalita, B., and Mazor, Z., Efficacy and safety of Glaubid® (Piloplex 3.4) in a long-term clinical study, presented at Int. Symp. on Glaucoma, Jerusalem, August 15, 1983.
18. Gurny, R., Latex systems, in *Topics in Pharmaceutical Sciences*, Breimer, D. D. and Speiser, P., Eds., Elsevier, Amsterdam, 1983, 277.
19. Gurny, R., Systèmes dispersés, in *Systèmes Thérapeutiques nouveaux et expérimentaux*, Puisieux, F., Rowland, X., and Buri, P., Eds., Technique et Documentation, Paris (in press).
20. Boye, T., Gurny, R., and Buri, P., Développement et évaluation d'un nouveau type de forme ophtalmique à libération prolongée, in Int. Symp. on Biopharmaceutics and Pharmacokinetics, Salamanca, Spain, 1984.

Chapter 5

MICROCAPSULES AND MICROSPHERES FOR EMBOLIZATION AND CHEMOEMBOLIZATION

J. P. Benoit and F. Puisieux

TABLE OF CONTENTS

I. INTRODUCTION

Pharmaceutical forms have shown meager developments in the past few decades, having been to some extent neglected in favor of the synthesis of new active molecules. Yet a major multi-disciplinary effort has been mounted in recent years to develop galenic forms and two research trends are currently being investigated.

The first is aimed at a closer understanding and control of the kinetics of drug release from the pharmaceutical form. This objective spurred the development of controlled release reservoir systems, especially by the Alza Company (Palo Alto, Calif.).[22,54] The second scheme calls for enhancing the specificity of medicinal products by routing the active ingredient to the organ, the tissue, or even the target cell.

A number of submicroscopic systems have been developed to attain this objective, such as liposomes[56] and nanospheres.[16] Microcapsules and microspheres for chemoembolization are pharmaceutical forms which highlight both trends in galenic research today: the controlled release of an active ingredient and the vectorization of the drug to the target organ.

This chapter examines the following aspects in succession:

1. Embolization and chemoembolization from a medical standpoint
2. Technological aspects pertaining to microcapsules and microspheres for chemoembolization
3. Investigations into this pharmaceutical form in the animal and man

II. EMBOLIZATION AND CHEMOEMBOLIZATION

A. Definition

Embolization is a therapeutic method designed to obliterate the vessels supplying a given pathological area by the endovascular pathway.[18] It is a relatively old technique. It was proposed by Brooks[11] in 1930, but only spread in France much later, on the initiative of Djindjian et al.[19]

B. Embolization Technique[18,41]

1. Anesthesia

Embolization is sometimes conducted under general anesthesia with intubation. As a rule, anesthesia with a patient awake (neuroleptanalgesia) or premedication with local anesthesia is preferably employed. The latter is routinely practiced in hepatic embolizations, in which the metabolism of the anesthetics administered slows down.

2. Angiographic Assessment

Embolization is always preceded by a complete angiographic assessment which serves to prepare a vascular chart of the area to be embolized. This helps to pinpoint the location of the feeder pedicles to be embolized.

3. Catheterization

The method most frequently used is the femoral pathway, which allows hyperselective catheterization in most cases. However, other paths can be considered (left axillary duct). A catheter is inserted into the previously identified pathological arteries. Use is generally made of a fine catheter (0.3 to 1 mm in diameter), flexible and hollow, of polyethylene or Teflon®, and adapted to the dimensions and curvature of the vessel scanned.

4. Practical Steps in Embolization

Once the catheter is inserted in the vessel to be embolized, the emboli are injected manually under slight pressure and in small volume groups. Each emboli injection is monitored by televised radioscopy by injecting a contrast material. Embolization is continued until a local circulatory slowdown is obtained, which is assessed according to the rate of clearance of the contrast material injected during examination.

The time required for embolization is normally quite long (2 hr or more).

5. Consequences of Embolization

After the operation, the patient must be hospitalized in an intensive care unit. The doctor may be led to:

- Correct a water electrolyte disorder due to the injection of an excessive quantity of contrast material and to parenchymatous ischemia
- Relieve pain in the embolized area, which subsides in 24 to 48 hr
- Treat a local edematous reaction (temporary intubation in case of embolization of the rhino-pharyngeal area)

C. Types of Emboli

A classification of the different emboli has been suggested by Kunstlinger et al.[33]

1. Particulate Emboli of Natural Origin

Use has been made of autologous clots, including unmodified fresh clots, fresh clots treated with heat or with substances such as ε-aminocaproic acid, or older clots. Irrespective of the type of clot used, it leads to lysis of the material, occurring very early, together with fragmentation, which is responsible for the distal migration of the emboli. The use of fragments of muscle and subcutaneous tissues taken from the patient himself has been reported. These emboli are not chemically resorbable and embolization is permanent.

Fragments of freeze-dried nonautologous dura mater (pig) are widely used in angiotherapy.

2. Particulate Emboli of Synthetic Origin

Cross-linked gelatin, marketed under the name Gelfoam® or Spongel® (ISH Laboratories), is a resorbable embolization agent. Easy to use, it can be made radio-opaque. Gelfoam® resorption is complete within 7 to 21 days. Oxidized cellulose (Oxycel®) and cellulose acetate (Surgicel®) have been proposed as emboli. They are resorbed in vivo within 6 to 8 weeks. Ivalon® occurs in the form of a compressible sponge of polyvinyl alcohol. It is a nonresorbable embolization material. Although it can be made radio-opaque, it is difficult to display in vivo. It is not toxic, but causes a slight inflammatory reaction.[32]

Spheres of lead and stainless steel form another group of synthetic materials. Lipid and polymeric microcapsules and microspheres constitute the final class of solid emboli. The technological, pharmacological, and clinical aspects of these microparticulate systems are dealt with in Sections III and IV.

3. Fluid Emboli

Fluid emboli are of three types:

1. Silicone prepolymers
2. Alkylcyanoacrylates
3. Suspensions of barium sulfate

Silicone prepolymers (polydimethylsiloxanes) are polymerizable *in situ* at ambient temperature in the presence of tin octanoate. Silicones do not induce any vascular reaction and are known to be completely nontoxic. Alkyl-α-cyanoacrylates are low viscosity liquid monomers. They polymerize in the vascular bed in contact with biological liquids. The polymerization rate is inversely proportional to the length of the alkyl chain. n-Butyl and isobutyl monomers, the only monomers used in man, polymerize less rapidly than methyl monomers. Polyalkylcyanoacrylates derived from n-butyl and isobutyl monomers are very slowly resorbable, if at all. No general toxicity or carcinogenesis has been observed. However, these results were obtained from relatively short studies. The absence of long-term experience imposes a minimum of skepticism towards these materials. Barium sulfate suspension, the third category of fluid emboli, are less widely used than the other types.

Apart from these emboli, the following deserve to be pointed out: (1) mechanical systems used for proximal obliteration (balloon sondes, metal spirals); and (2) electrocoagulation, which helps to obliterate the vascular bed in a pathological area.

D. Goals of Embolization

As a rule, the embolization of a lesion area has a number of elementary effects.[34]

1. Antalgic Effect

The occurrence of pain can, for example, be related to the mass of a tumor (hepatoma, metastasis in the liver).[60] By favoring necrosis of the tissues, embolization leads to a reduction or durable disappearance of algic phenomena in most cases.

2. Antihemorrhagic Effect

Embolization helps to achieve hemostasis by *in situ* occlusion of the pedicles responsible for spontaneous or preoperative bleeding. In digestive pathology, in the case of gastric and duodenal ulcers, embolization helps to solve hemorrhage problems.[60]

3. Effect on the Susceptibility to Change of a Benign Lesion

Indirectly, embolization allows the total ablation of a lesion which has recurred on a postsurgical residue. Moreover, it directly causes ischemic necrosis of the lesion. Therapeutic carcinological action can be undertaken by this technique. A number of spectacular results have been obtained with hepatomas.[60]

4. Antisecretory Effect

This effect also results from the ischemia and necrosis of a pathological hormone tissue. In the case of nonchromaffin paragangliomas, embolization has helped to stem the secretory activity of the pathological tissues (serotonin or catecholamine).[34]

E. Indications of Embolization

Therapeutic embolization can be proposed in three ways: preoperative, palliative, and curative.[18]

1. Preoperative Embolization

The preoperative embolization of the arterial pedicles of a hypervascularized and extensive tumor enables the surgeon to carry out his operation more completely and rapidly. As a rule, embolization reduces preoperative hemorrhage losses. In addition, due to a perilesional edema and a tumoral reduction secondary to ischemic necrosis, exeresis of the tumor is easier.

For example, a large nasopharyngeal angiofibroma was extracted after embolization of the different tumoral arterial pedicles using dura mater, a solution of Thrombine®, and Hemocaprol®.[18] Meningiomas, which are benign but richly vascularized tumors, may be dangerous to operate on because of their volume or location. The obliteration of the arterial feeder pedicles crossing the base of the skull serves to eliminate any risk of hemorrhage during the operation, and helps to drain and to necrose the tumor, which becomes more accessible to the surgeon.[44]

2. Palliative Embolization

The largest number of embolization indications fall in this group. If an operation is unfeasible, palliative embolization may be carried out for an antalgic or, above all, a hemostatic purpose, after the failure of medical treatment.

In pathological cases in which any surgical act is contraindicated (alteration of the patient's respiratory function, wide extension of a lesion), embolization is the only possible treatment. The treatment of serious hemoptysis by this method is a good example.[10]

3. Curative Embolization

The devascularization achieved by embolization may be distal and selective on limited vascular lesions, contrary to surgery. The tissue after-effects secondary to the resulting ischemia are therefore narrowly localized. In certain cases, the final and stable result obtained after embolization helps to avoid an operation. A good example of this indication is the hyperselective embolization of traumatic renal arterial venous fistulas, as opposed to and at best partial, but often total nephrectomy of the affected kidney.[18]

Similarly, the endovascular treatment of arteriovenous, carotid-cavernous, and vertebro-vertebral fistulas, which are inaccessible to neurosurgical treatment, belong to the group of curative embolizations.[44]

F. Contraindications of Embolization

These are of three types.[18]

1. Technical Contraindications

The unfeasibility of hyperselective catheterization may be related to a specific anatomical configuration or an obstruction of the advance of the catheter (atherome).

2. Anatomical Contraindications

Embolization is contraindicated if the arterial pedicle to be obliterated gives rise to a "noble" artery. Another case is the possible embolization of a large arteriovenous angioma or a high output arteriovenous fistula. At any rate, the injection of fragmented emboli is contraindicated because of the risk of pulmonary embolism if the embolus reaches the systemic circulation.

3. Clinical Contraindications

These are few in number. The main contraindication is a poor indication. This is because, while embolization is a reliable technique, it incurs a number of risks to which it is useless to expose tired patients if the final result is irrelevant. Its indication must therefore be discussed by the team of doctors, surgeons, reanimators, and radiologists who bear responsibility for the case.

G. Chemoembolization

1. Definition

Whereas embolization aims in general at three types of ailment (hemorrhages, vascular malformations, and tumors), chemoembolization has only one field of action and may be considered as a new approach to antimitotic treatment in oncology. Chemoembolization is completely identical to embolization from the technical standpoint, inasmuch as the occlusive agents are vectors of an antimitotic drug. At the present time, the available emboli are either microcapsules[24,30] or microspheres[5,55,63,65] containing an antitumor drug.

2. Advantages of Chemoembolization in Comparison with Classic Anticancer Treatment

According to Kato et al., this technique offers the following advantages:[28-30]

1. The presence of an effective concentration of active ingredient for a longer interval in the target tissue
2. A decrease in side effects associated with the location of the active agent within the tumor and with lower plasma concentrations in the peripheral blood
3. A reduction in the total quantity of drug to be administered
4. Potentialization of the antitumor activity of the active ingredient by the mechanical effect of obliteration, which causes hypoxia of the embolized tissue

This chemoembolization technique is feasible whenever an artery irrigates a localized tumor area.

III. MICROCAPSULES AND MICROSPHERES

A. Definition

1. Microcapsules and Microspheres for Embolization

The French Pharmacopée[53] offers a general definition of microcapsules as solid materials consisting of a solid envelope containing a liquid, a solid, or a pasty substance. They are produced by various methods, including coacervation and interfacial polymerization. Microcapsules occur in the form of a powder with particles less than 1250 μm in diameter. This definition applies equally to microcapsules for embolization.

The French Pharmacopée does not offer a definition for microspheres. By analogy with the foregoing definition, microspheres may be considered as "full" solid and spherical materials with a particle size range on the same order of magnitude as that of microcapsules.

2. Microcapsules and Microspheres for Chemoembolization

The chemotherapeutic effect is provided by the presence of an antitumor substance in these microparticulate systems. The liquid, solid, or paste content of the microcapsules consists of molecules of active substance in the solid or liquid state.

The internal phase of microspheres consists of molecules of active substance dispersed or dissolved in the material making up the particles. Each microsphere is actually a matrix in which the active ingredient is disseminated.

The preparation of microspheres and microcapsules requires the simultaneous presence of the support material and the active molecule in the reaction medium. However, they involve different technologies, and coacervation, for example, is inapplicable for microsphere manufacture.

B. Raw Materials

1. Support Materials

The materials selected must meet certain requirements:[62]

1. Ability to be shaped to the desired form
2. Release of the active substance by an appropriate mechanism
3. Stability of the form and adequate mechanical strength
4. Nontoxicity of the system itself and of any of its degradation products
5. Bioinertia or biodegradation
6. Sterilizability

The support materials selected in our laboratory belong to two main classes:

- Nonpolymeric compounds
- Polymeric compounds

The first class is represented by two lipids, carnauba wax[58,63] and cetyl alcohol.[65] Carnauba wax, which essentially consists of fatty acid esters, is a wax obtained from the leaves of the palm tree *Copernicia cerifera.* It was selected chiefly for its good biocompatibility, biostability, and normally ideal melting point (82 to 86°C).

The French Pharmacopée describes officinal cetyl alcohol as a mixture of solid saturated alcohols, represented mainly by hexadecanol (Figure 1). Cetyl alcohol was selected as the support material because of its low melting point (50°C) which allows the incorporation of thermosensitive active substances.

These two lipids, carnauba wax and cetyl alcohol, are used in the preparation of microspheres for chemoembolization.

The second class of support material consists of natural and synthetic polymers. The natural polymers employed are biodegradable. They include human serum/albumin cross-linked with terephthaloyl dichloride.[6] This material is suitable for both microspheres and microcapsules. The degree of cross-linking of the final protein material serves to adjust its biodegradability.

A polyester synthetized by a bacteria is currently under investigation,[9] poly(β-hydroxybutyric acid). Only microspheres have been prepared with this material. The synthetic polymers investigated are essentially biodegradable polyesters derived from lactic acid and glycolic acid.[3-5] Their biodegradability is closely related to their chemical composition.[66,67] Hence, the homopolymer poly(D,L-lactic acid) biodegrades in 6 months.[45] Biodegradability can be accelerated by increasing the proportional glycolic units in this polymer. For example, a 50/50 poly(lactic acid, glycolic acid) polymer biodegrades in 1 week.[45]

A nonbiodegradable hemisynthetic polymer, ethylcellulose, has been used to prepare microcapsules for chemoembolization.[24]

This material is certainly the one that is the most widely used as a microcapsule wall material. Kato et al. use this type of material exclusively to prepare microcapsules for chemoembolization.

2. Active Ingredients

Two antitumor substances conveyed by microparticles have mainly been used in the laboratory (Figure 2): (1) 5-fluorouracil, which is strongly hydrophilic; and (2) lomustine, or CCNU (1-(2-chloroethyl)-3-cyclohexyl-1-nitrosourea), which is lipophilic. These two molecules are selected in accordance with two criteria, one physicochemical and the second medical.

The extreme hydrophilic and lipophilic properties of the active molecules raise diametrically opposed galenic problems which, once resolved, allow the encapsulation of any antitumor substance with intermediate hydrophilic properties.

These two molecules are highly interesting from the therapeutic standpoint. 5-Fluorouracil inhibits the biosynthesis of nucleic acids.[35] To be active, 5-fluorouracil must be converted to 5-fluorodeoxyuridine-monophosphate (FdUMP), responsible for its cytotoxicity. In fact, FdUMP specifically inhibits thymidilate synthesis. Blockage of the synthesis of thymidine monophosphate causes the stoppage of DNA synthesis and consequently the death of the cell.[35] 5-Fluorouracil, alone or combined with other antitumor substances, is chiefly prescribed for solid tumors of digestive and hepatic origin, also for tumors of the kidney and the ovary.

FIGURE 1. Chemical structure of hexadecanol.

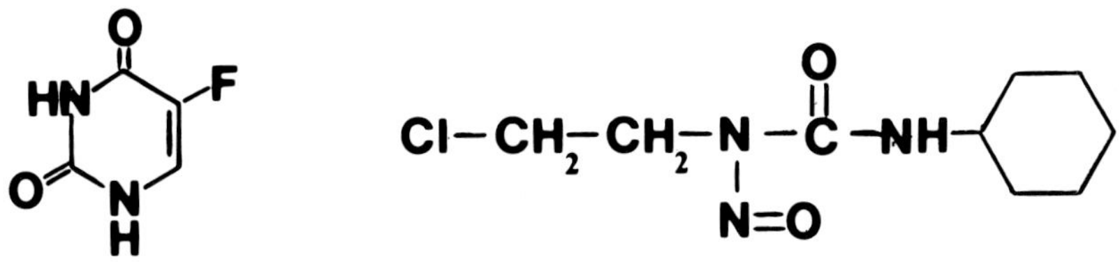

(a) (b)

FIGURE 2. Chemical structures of 5-fluorouracil (a) and lomustine (b).

Lomustine belongs to the class of nitrosoureas known as alkylating agents. Although their action mechanism is not thoroughly understood, many investigations into the behavior of nitrosoureas show that they decompose rapidly in physiological conditions to give rise to highly reactive entities.[46,47,59] The carbonium ions generated by the decomposition of CCNU interact with the nucleophilic groups of nucleic acids. This causes their cross-linking as shown by Figure 3.[32]

Lomustine is of special interest in the field of cytostatics, due to its lipophility, which enables it to cross the hematoencephalic barrier. Consequently, its prime indication is for the treatment of tumors of the central nervous system.[12] It also displays chemical activity against lymphomas, malignant melanomas, and gastrointestinal carcinomas.[12]

Kato et al. use mitomycin C as an active substance (Figure 4).[30] This water-soluble molecule is an antitumor antibiotic extracted from *Streptomyces caespitosus*.[17] The molecule has an aziridine nucleus and a carbamate group which, once activated by in vivo reduction of the benzoquinone nucleus to the semiquinone radical, forms covalent bonds with the guanine and cytosine bases of DNA.[17] This inhibits the synthesis of DNA.

The main indications of mitomycin C are[17]

1. Tumors of the digestive, urinary, or ORL areas
2. Primary and secondary bronchopulmonary cancers
3. Mammary adenocarcinomas
4. Chronic myeloid leukemias

It should be emphasized that this molecule is highly toxic, especially from the hematological standpoint, resulting in thrombopenia. In repeated doses, it also causes immunodepression.[21]

C. Preparative Methods

Preparative techniques are among those routinely used for microencapsulation.

1. Preparation of Microspheres

The microspheres are prepared in the laboratory by three different techniques: (1) hot melt of the support material; (2) solvent evaporation; and (3) stalagmopoiesis.

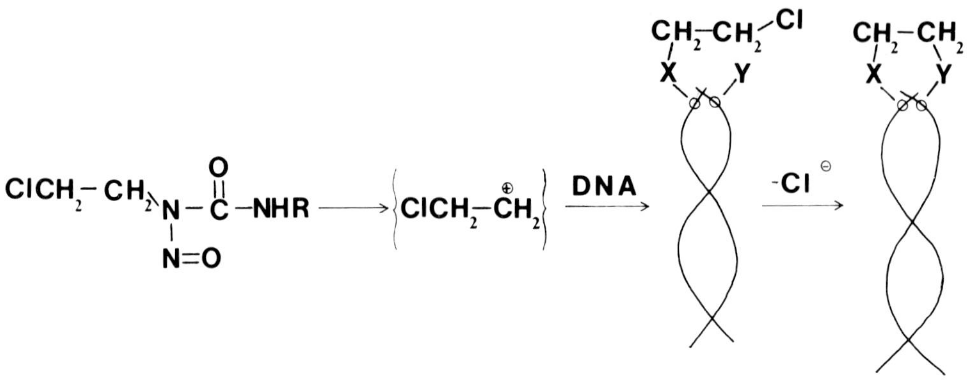

FIGURE 3. Mechanism proposed for the cross-linking of DNA by a *N*-(-2-chloroethyl)-*N*-nitrosourea.[32] X and Y = nucleophilic sites on two opposite DNA strands.

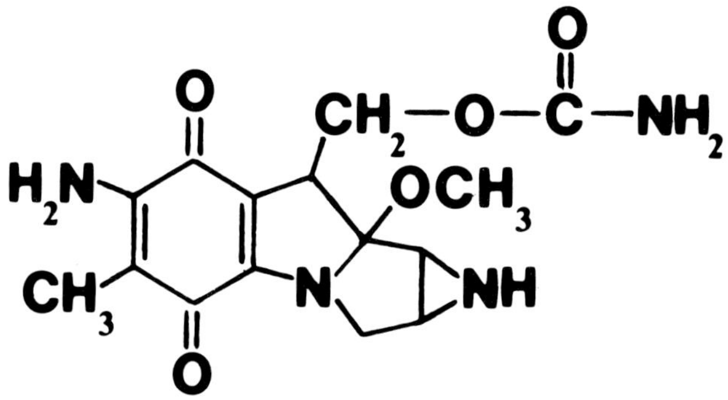

FIGURE 4. Chemical structure of mitomycin C.

a. *Hot Melt of the Support Material*

This method involves 4 steps (Figure 5):[55]

1. The support material is fused just above its melting point
2. The active ingredient is dispersed in the molten material
3. The molten support/active ingredient combination is emulsified in a suitable dispersant phase such as silicone oil or distilled water
4. The dispersed globules are solidified by adding to the dispersion a certain quantity of dispersant phase cooled to low temperature

The microspheres are separated by filtration, washed, dried, and sieved to separate the particle size fractions. The particle size distribution is an important factor because it conditions the location of the microspheres in the arterial system. It can be adjusted to the needs of the medical profession by altering technological parameters such as the speed of agitation and the viscosity of the dispersant phase.

This so-called hot melt technique is reserved for lipid materials such as carnauba wax and cetyl alcohol, whose melting points are not very high.

b. *Solvent Evaporation*

This method, used by Thies,[64] is illustrated in Figure 6. The active molecule to be encapsulated and the microsphere wall material are dissolved in a volatile organic sol-

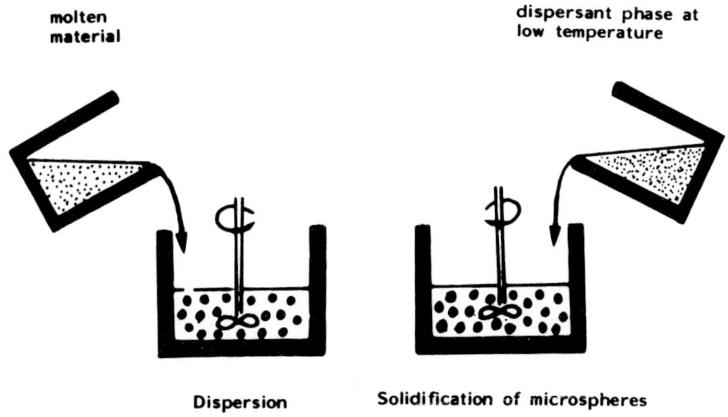

FIGURE 5. Microspheres, preparation by hot melt of the support material.[55]

vent which is immiscible in water. The resulting solution is emulsified in an aqueous phase containing a surface active agent. The organic solvent gradually evaporates with agitation and the microspheres are thus formed.

Although the concept of this process is very simple, the final result is affected by many variables: nature of the organic phase, volume proportions of the organic and aqueous phases in contact, evaporation temperature, quantity of support material, type and amount of surface active agent, and solubility of the drug to be encapsulated.

This method is ideal for the preparation of microspheres of polymers such as poly(D,L-lactic acid).

c. Stalagmopoiesis[55]

This preparative method uses an apparatus developed by Ross et al.[61] (Figure 7). The molten material produces drops in nozzles, and the drops solidify as they fall through a volume of air. Used for lipid materials, this technique allows the preparation of large quantities of microspheres characterized by a relatively narrow particle size distribution.

2. Preparation of Microcapsules

The most widely used method for chemoembolization is the coacervation of a polymer such as ethylcellulose, by lowering the temperature. Figure 8 illustrates the coacervation mechanism. Above the phase separation curve (point A), a homogeneous solution of polymer constitutes the reaction medium. When the temperature of the medium is lowered, the phase separation curve is reached (point E). At a lower temperature than the phase separation temperature T_c, the reaction medium separates into two phases (point B): one phase rich in solvent and poor in polymer (point C); the second poor in solvent and rich in polymer, the coacervate phase (point D).

This phase deposits progressively on the drug particles in suspension in the initial reaction medium (Figure 9). The isolation of the microcapsules by filtration requires the solidification of the coacervated wall of the microcapsules by methods such as the application of cold or desolvation.

Microcapsules of 5-FU were prepared by this technique in our laboratory, using the ethylcellulose/cyclohexane system.[24] Kato's Japanese team exploited this method to prepare microcapsules of mitomycin C with the same polymer/solvent system (ethylcellulose/cyclohexane).[25]

A second method is employed in the laboratory to prepare biodegradable microcapsules. It is based on interfacial polymerization, investigated mainly by Chang.[14,35] An

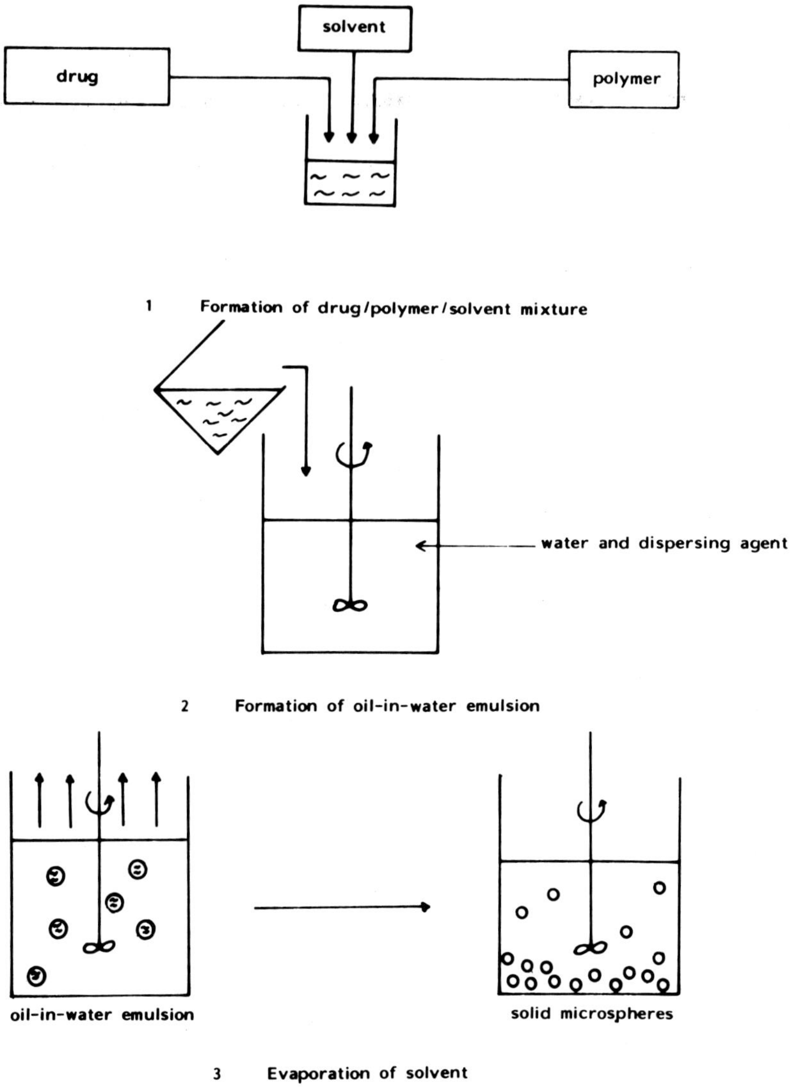

FIGURE 6. Microspheres, preparation by solvent evaporation.[64]

aqueous solution of human serum/albumin is emulsified in a cyclohexane/chloroform (4/1) mixture. A difunctional acid chloride such as terephthaloyl chloride is added to the organic phase. This chemical agent cross-links the protein by reacting with the amine groups (Figure 10). This microcapsule preparation method is important inasmuch as the degree of cross-linking and hence the biodegradation can be adjusted in accordance with certain parameters such as the quantity of acyl dichloride.[6] The anti-tumor substance is incorporated in this type of microcapsule by extemporaneous impregnation employing a solution of active ingredient.

D. Sterilization and Presentation

Microspheres are sterilized in two different ways: by ethylene oxide or by ionizing radiation.

Sterilization by ethylene oxide is carried out for 24 hr at 45°C. Desorption is then performed for 2 months in a ventilated room at ambient temperature. Apart from its

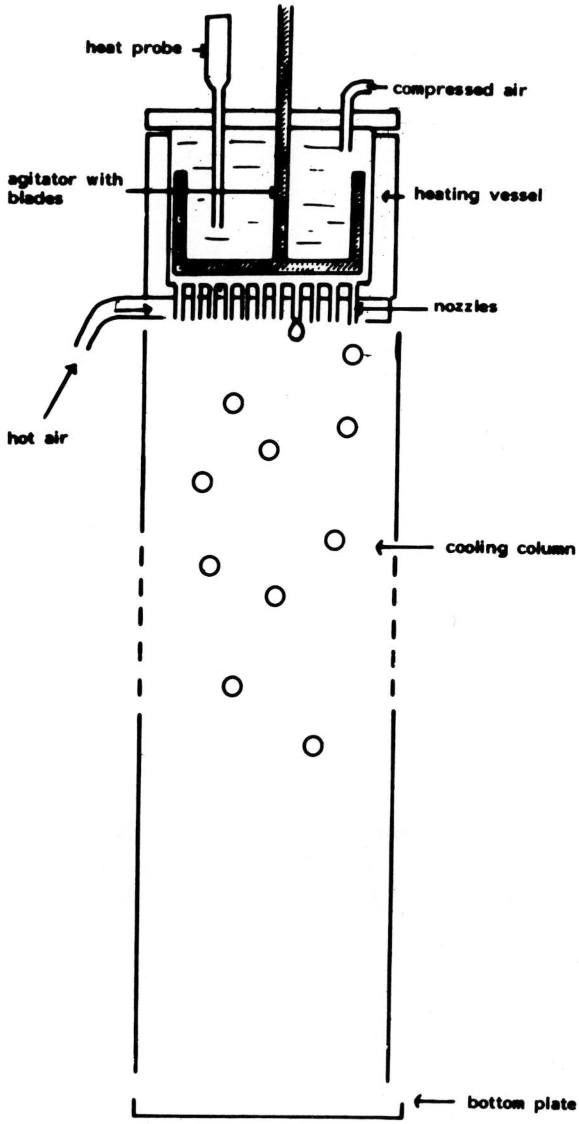

FIGURE 7. Microspheres, stalagmopoiesis.[61]

length, this sterilization method presents the drawback of leaving high residual concentrations of ethylene oxide in certain materials. For example, the presence of ethylene oxide concentrations of about 45 ppm has been demonstrated in lomustine loaded cetyl alcohol microspheres.[65] The French Pharmacopée specifies a maximum concentration of 2 ppm.[51] For these reasons, this sterilization method has been abandoned in favor of radiosterilization.

Sterilization by ionizing radiation is carried out using an electron accelerator which serves to focus an electron beam on the microparticles. The ionizing radiation dose involved is 2.5 Mrad,[52] the minimum dose indicated in the French Pharmacopée.

An investigation on microspheres of cetyl alcohol, which are vectors of lomustine, a particularly labile molecule, showed that the dose employed failed to cause any physical or chemical alteration of the particles. This sterilization method has been extended to all the microparticulate systems prepared in the laboratory.

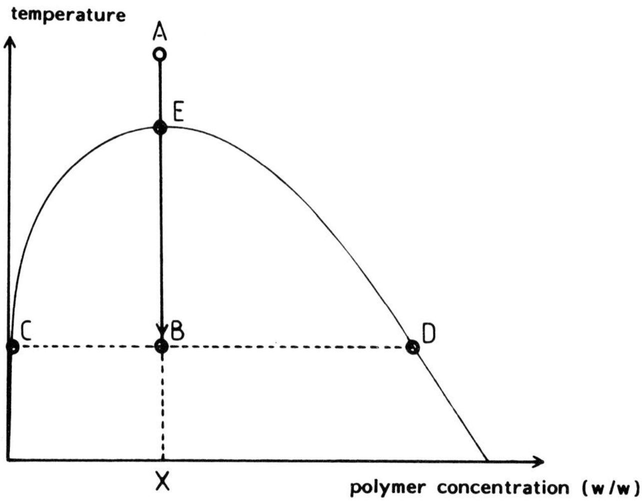

FIGURE 8. Microcapsules, coacervation by temperature lowering, principle.

Particles weighing 100 to 250 mg are distributed in glass bottles. One bottle containing 30 mℓ of sterile aqueous saline solution with 0.1% Pluronic F68® serves to make a fresh suspension of the microspheres. Certain microspheres are placed directly in suspension in an aqueous solution of an iodized contrast material. In their study, Kato et al. sterilized their mitomycin C microcapsules at 140°C for 1 hr.[29]

E. Characterization of Microspheres and Microcapsules for Chemoembolization

A number of characterizations must be performed on microparticulate systems. The physicochemical characteristics determined may affect their aptitude for use as emboli. This is illustrated by presenting the characterizations made on microspheres of poly(D,L-lactic acid) with and without active ingredient.[5,57] Three active substances were incorporated: lomustine, an antitumor substance and two substances selected as tracers, progesterone and hydrocortisone.

1. Morphological Analysis

This study was conducted by scanning and transmission electron microscopy. In nonloaded or slightly drug-loaded (23%) microspheres, the surface of the particles is perfectly homogeneous and smooth in appearance. No pores or crystals are visible (Figure 11). As the drug content is increased to a moderately high level of 35%, structural changes occur. A few active ingredient crystals appear at the surface embedded in the polymer matrix. The external surface is not as smooth as before (Figure 12). With a very high drug content (68%), the particles assume a completely different appearance. While they are still spherical, the external structure consists of a continuous layer of active ingredient crystals and the microspheres become rougher in texture (Figure 13).

The scanning electron microscope also helps to determine the surface texture of the microspheres. This may have a significant physiological impact. Microspheres prepared in our laboratory and tested on animals present the drawback of migrating distally in the arterial system due to a reaction vasoplegia caused by embolization. These microspheres [carnauba wax and poly(D,L-lactic acid)] display a smooth surface. Particles with an irregular, rough surface are liable to attack the vascular endothelium. This would cause the formation of a thrombus that would block the microspheres at the desired location by a complementary obliteration mechanism.

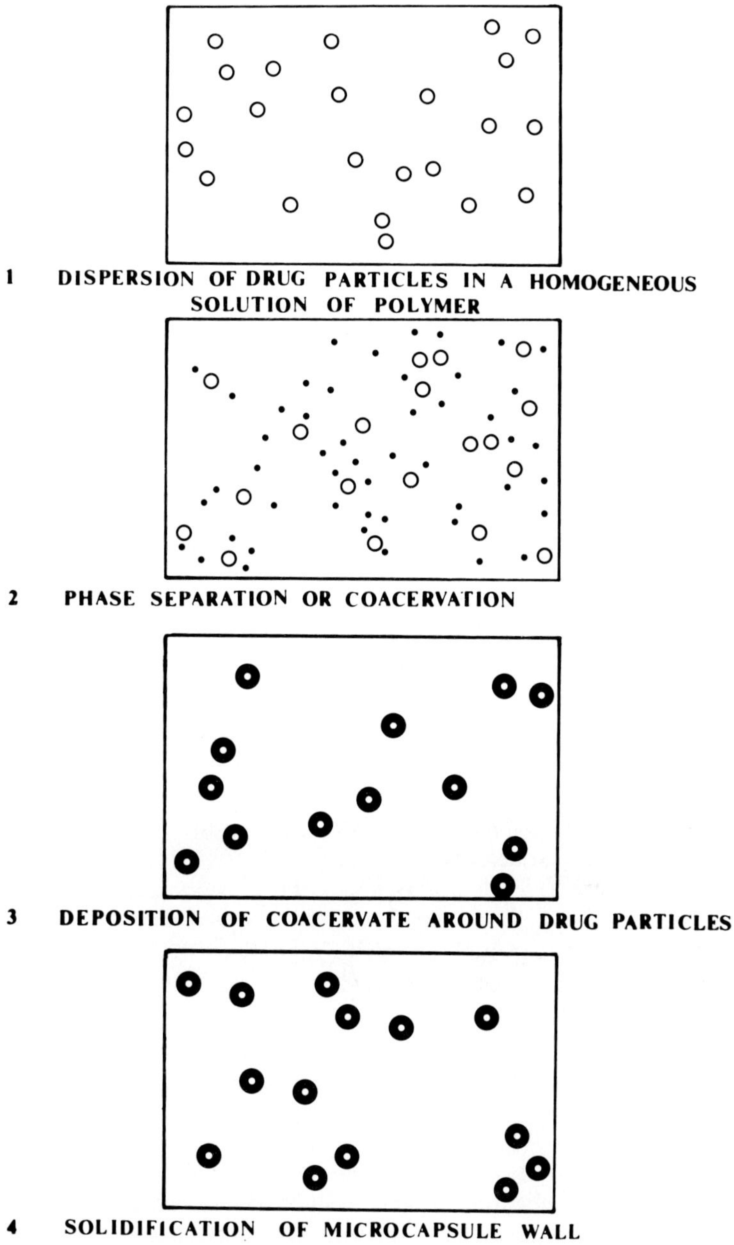

FIGURE 9. Ethylcellulose microcapsules, preparation by coacervation.

2. Particle Size Analysis

This analysis is indispensable because the size of the particles conditions their intra-arterial location. Various methods are available, such as the electronic particle counter (Coulter Counter®).

During the preparation of the microspheres, technological factors affect the particle size distribution. These include the speed of agitation, viscosity of the reaction medium, and the presence of an active substance. Figure 14 illustrates the influence of speed of agitation of the medium on the average diameter of the microspheres of poly(D,L-lactic acid) loaded with lomustine and nonloaded.

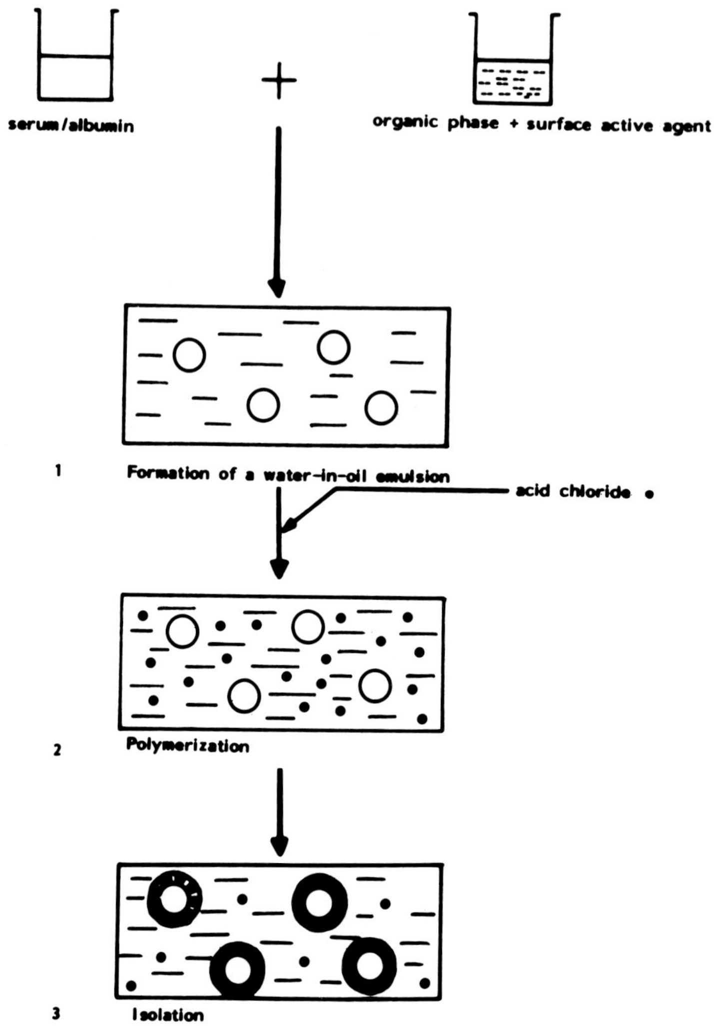

FIGURE 10. Cross-linked serum/albumin microcapsules, preparation by interfacial polymerization.[6]

3. Internal Structural Analysis

This study helps to determine the physical state of the encapsulated anti-tumor drug, which depends on many factors. This state is important to determine in so far as it affects certain major properties, including the drug release rate and the stability of the finished medicinal product.

Microspheres of poly(D,L-lactic acid), taken as an example in this chapter, were subjected to a study for lomustine and also for two other active substances used as tracers (progesterone and hydrocortisone). These two substances were used to evaluate the influence of the lipophility of the active substance in its physical state in the microsphere. The technique is based on differential thermal analysis (DTA).

a. Lomustine Microspheres

Figure 15 illustrates the thermograms of poly(D,L-lactic acid), nonencapsulated lomustine, and lomustine microspheres (23%). The plot of the support polymer displays a phase transition at 57°C, corresponding to the glass transition temperature. Below

FIGURE 11. Microspheres of poly(D,L-lactic acid) containing 23% lomustine, scanning electron microscope. Magnification × 100.[5]

this temperature (T_g), the polymeric material is hard and brittle, but displays considerable elasticity above this temperature.[13]

On the molecular level, glass transition corresponds to a change in the degrees of freedom of the movements of macromolecular chains located in the amorphous areas of the polymer. For nonencapsulated lomustine, the DTA plot shows a clear endotherm at 94°C, corresponding to the melting of the cytotoxic crystals. This transition is accompanied by exothermic signals which can be attributed to the decomposition of the lomustine after fusion.

The thermograph of the lomustine microspheres reveals three essential elements. The inflection at 57°C characterizing the glass transition of the polymer disappears and is supplanted by a broadening of the plot which extends from 20 to 100°C. The glass transition temperature is less clear and is displaced towards lower temperatures. It was found to be equal to 20°C by differential scanning calorimetry.

At 94°C, the thermogram does not display the signal corresponding to the fusion of lomustine. Exothermic signals continue to occur around 110°C. These results serve to clarify the physical state of encapsulated lomustine (Figure 16).

The disappearance of the peak corresponding to the fusion of drug crystals clearly proves the absence of lomustine in the crystalline state in the microspheres. The structural hypothesis of crystalline dispersion can therefore be discarded. According to the classification of Chiou and Riegelman,[15] two other alternatives are available.

The first is that of molecular dispersion in a vitreous medium (glass solution). The active ingredient is accordingly dispersed in a metastable amorphous state in a polymer in the vitreous state. According to Chiou and Riegelman, crystallization of the dispersed product at the time of solvent evaporation can be prevented in such systems by

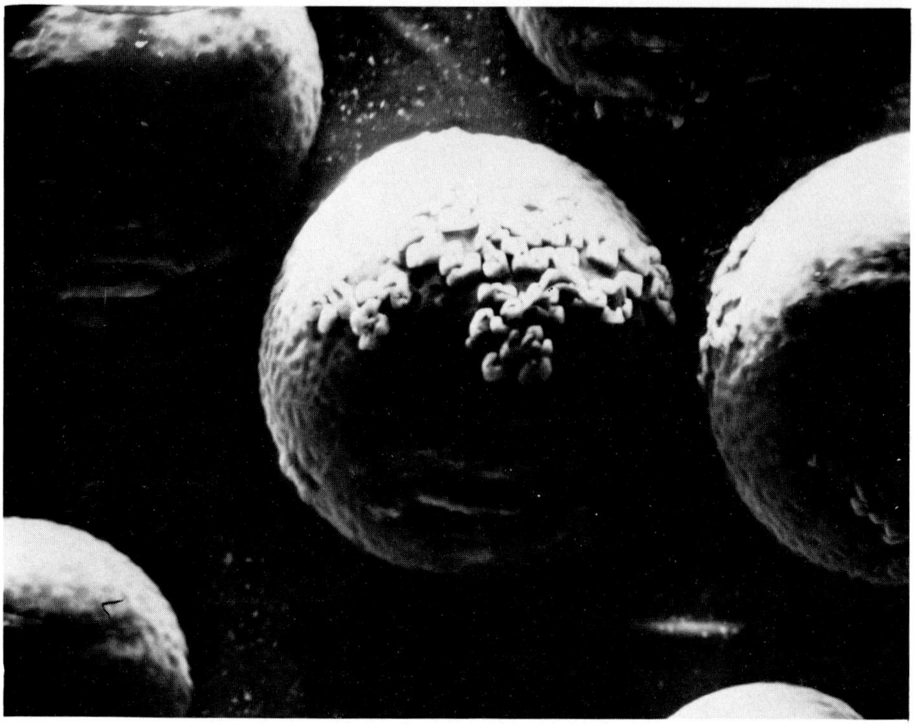

FIGURE 12. Microspheres of poly(D,L-lactic acid) containing 35% progesterone, scanning electron microscope. Magnification × 400.[5]

the high viscosity of the medium. For such a structure, the thermogram should not display a signal related to the fusion of the crystalline regions, but, on the other hand, no change in the glass transition temperature should be recorded. However, the thermogram of lomustine microspheres displays a shift in the T_g. Hence, the molecular dispersion in the vitreous medium is not the physical structure corresponding to encapsulated lomustine.

The second possibility is that of a solid solution. The lomustine molecules are accordingly embedded between the macromolecular chains. By separating them, they would tend to reduce the secondary polymer/polymer bonds. The free volume available for each chain and hence its degree of freedom, would accordingly increase. The lomustine would act as a sort of plasticizer for the polymer and apart from the absence of crystalline regions in the polymer mass, the glass transition temperature of the latter would be lowered.

This is exactly what is observed in actual fact (Figure 15). Lomustine hence appears clearly to be in the state of a solid solution in poly(D,L-lactic acid).

b. Microspheres of Progesterone and Hydrocortisone

To extend these results to other areas, two other active ingredients were investigated, the lipophilic molecule progesterone and the far more hydrophilic hydrocortisone. Their physical state in microspheres of poly(D,L-lactic acid) was determined by the method presented.

Progesterone — Figure 17 shows the overall DTA results obtained for progesterone. In the thermogram of progesterone microspheres (23%) two elements are worth noting: the endotherm of drug crystal fusion is absent; and glass transition of the poly(D,L-

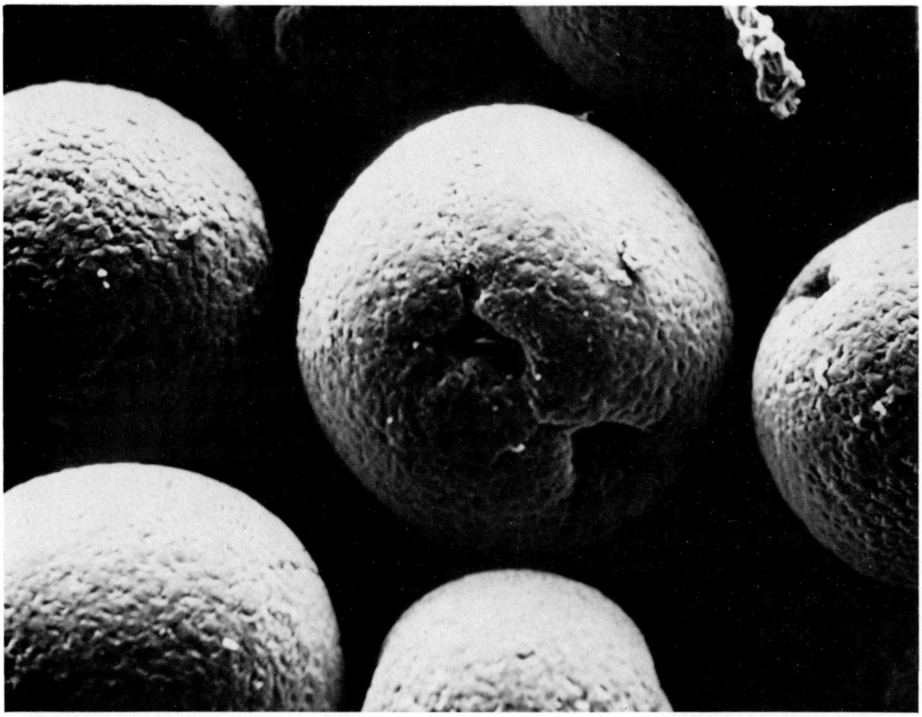

FIGURE 13. Microspheres of poly(D,L-lactic acid) containing 68% progesterone, scanning electron microscope. Magnification × 100.[5]

lactic acid) remains located at 57°C. Hence, the progesterone does not exist in the state of crystalline dispersion in the polymer. Nor is it in the state of a solid solution, since it does not display plasticizing properties towards the poly(D,L-lactic acid). Its probable state in the microsphere is that of a molecular dispersion in a vitreous medium.

This was confirmed by annealing experiments (Figure 18). During these experiments and before DTA the microspheres were raised to a temperature above T_g for a given interval. The thermogram obtained after annealing for 2 days at 80°C displays a signal at 125°C corresponding to the fusion of progesterone crystals. During annealing, the degree of mobility of the polymer chains increases (temperature above T_g) and the progesterone molecules can diffuse in the matrix and crystallize. This result confirms the hypothesis according to which the internal structure of the microspheres is that of a metastable molecular dispersion of active ingredient in the polymeric mass.

Hydrocortisone — The profile of the thermograms recorded for hydrocortisone alone and encapsulated (20%) is shown in Figure 19. For the crystallized active ingredient, the melting point appears in the form of a signal at 222°C. After encapsulation, this signal persists, albeit diminished, indicating that part of the hydrocortisone is dispersed in the microspheres in the form of crystals. This result seems logical in so far as, contrary to progesterone and lomustine, which are lipophilic drugs, hydrocortisone is insoluble in the methylene chloride used as an organic phase in preparation. The drug, encapsulated by the solvent evaporation process, is a suspension of hydrocortisone in methylene chloride.

c. Conclusion

Considering the results concerning lomustine, progesterone, and hydrocortisone microspheres, it clearly emerges that the internal structure of the microspheres depends

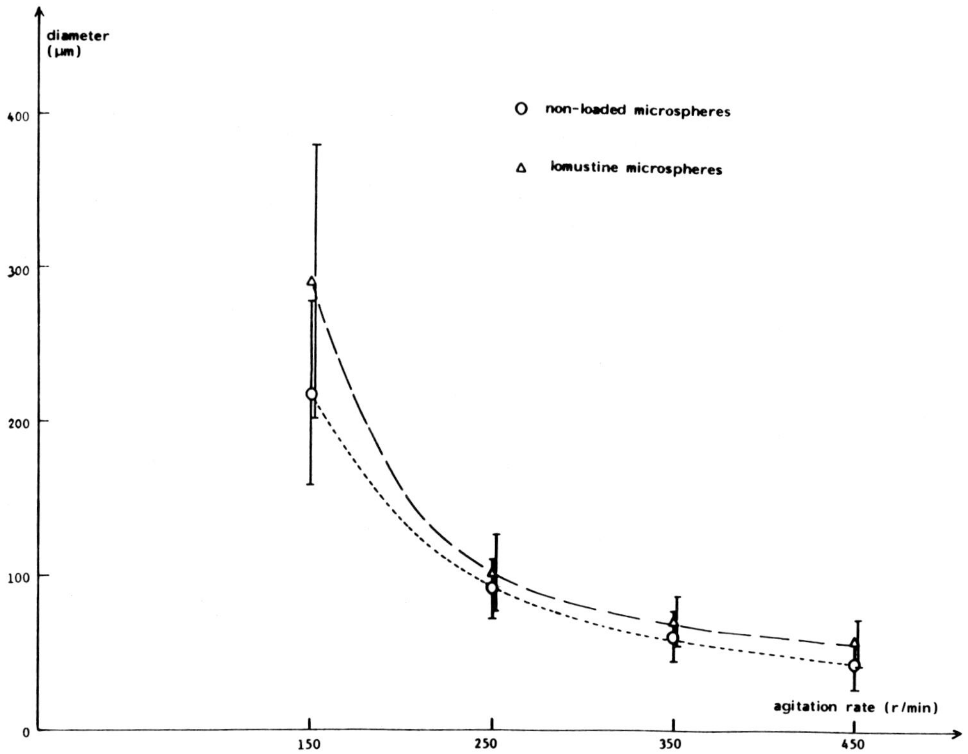

FIGURE 14. Microspheres of poly(D,L-lactic acid), effect of speed of agitation on microsphere diameter.[5]

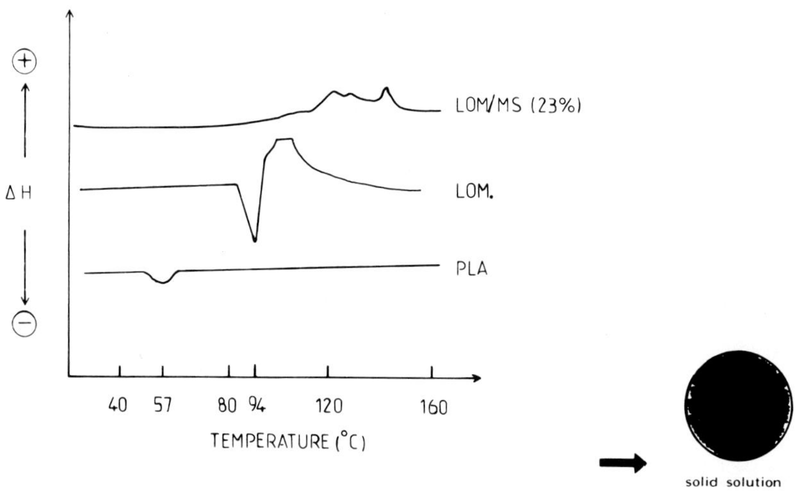

FIGURE 15. Microspheres of poly(D,L-lactic acid), thermograms of polymer (PLA), nonencapsulated lomustine (LOM), and lomustine microspheres (LOM/MS).[5]

on the lipophilic or hydrophilic properties of the medicinal agent (Table 1). If the drug is lipophilic (lomustine and progesterone), it is found either in the form of a solid solution (lomustine) or in the form of a metastable molecular dispersion in a vitreous medium (progesterone). In no case is the presence of crystals in the matrix detectable for an initial drug content of about 23%. On the other hand, if the active ingredient is

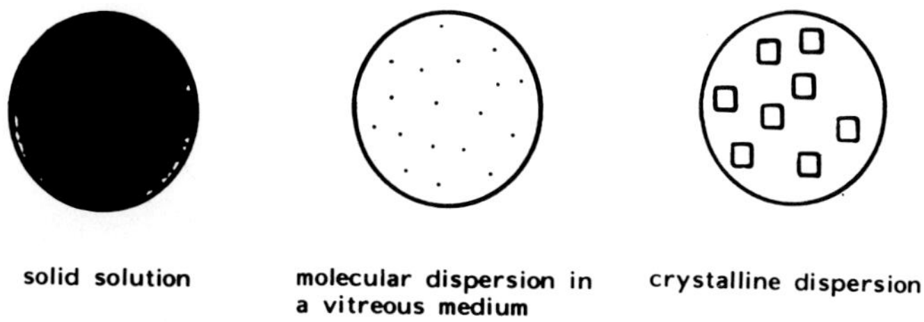

FIGURE 16. Microspheres, physical states of encapsulated drug.[5]

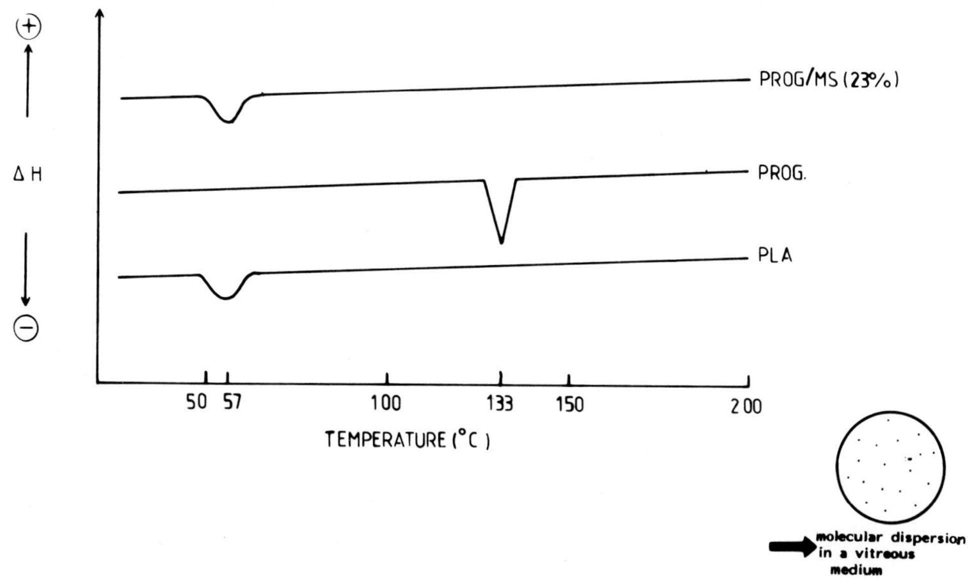

FIGURE 17. Microspheres of poly(D,L-lactic acid), thermograms of polymer (PLA), nonencapsulated progesterone (PROG), and progesterone microspheres (PROG/MS).[5]

more hydrophilic (hydrocortisone), it is partly dispersed in the form of crystals in the polymeric matrix.

From the standpoint of chemoembolization, this characterization is important because the physical state of the encapsulated antitumor substance affects its rate of release into the vascular bed. Moreover, the physical structure of the active molecule conditions its stability in the microparticles.

4. Quantitative Analysis

The rate of incorporation in microspheres is decisive for the quantity of active substance that can be conveyed to the target. In the case of microspheres of poly (D,L-lactic acid) prepared by solvent evaporation, this rate depends on the organic solvent/water partition coefficient of the active ingredient. Table 2 offers an example of the rates obtained with lomustine and the two tracers progesterone and hydrocortisone. This determination was conducted by HPLC.

F. Study of In Vitro Drug Release from Microspheres

One of the objectives of galenic research today is optimum control of the kinetics of drug release from the administered form. In the specific case of chemoembolization,

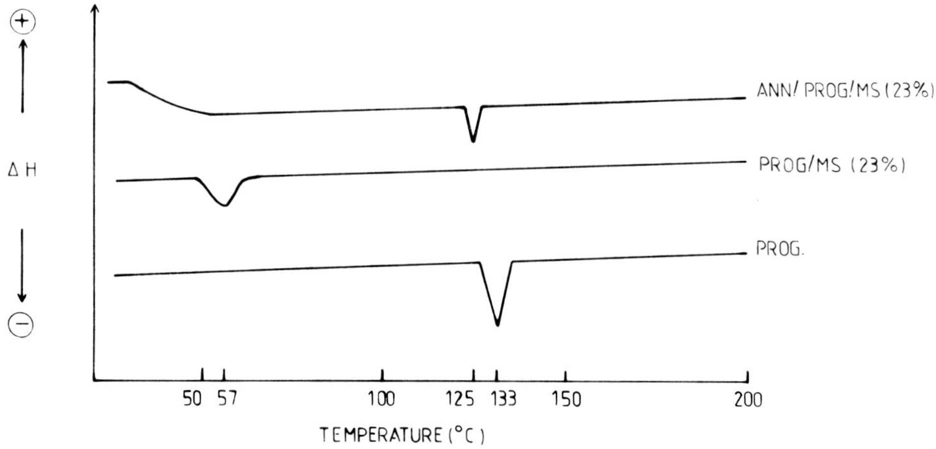

FIGURE 18. Microspheres of poly(D,L-lactic acid), thermograms of nonencapsulated progesterone (PROG), progesterone microspheres (PROG/MS), and annealed progesterone microspheres (ANN/PROG/MS).[5]

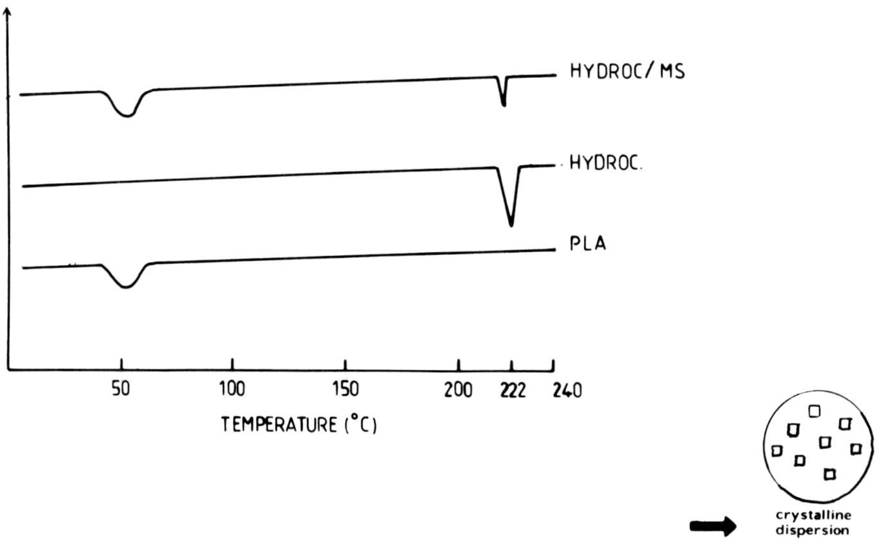

FIGURE 19. Microspheres of poly(D,L-lactic acid), thermograms of polymer (PLA), nonencapsulated hydrocortisone (HC), and hydrocortisone microspheres (HC/MS).[5]

the objective must be to offer the medical profession microemboli which release the active substance according to a kinetics that is suitably predictable in advance.

The kinetics of release of the active substance in the case at hand (microcapsules and microspheres) depend on the structure of the galenic form. For a microcapsule bounded by a nonporous polymeric form, in which the internal drug concentration can be considered as constant, the rate of release is given[1] by Equation 1:

$$\frac{dM_t}{dt} = \frac{A \cdot D \cdot K \cdot \Delta C}{l} \tag{1}$$

Table 1
MICROSPHERES OF POLY(D,L-LACTIC ACID), INFLUENCE OF THE PHYSICOCHEMICAL PROPERTIES OF THE ACTIVE INGREDIENT ON ITS PHYSICAL STATE IN THE MICROSPHERE[5]

Active ingredient	Physicochemical property	Physical state in microsphere
Lomustine	Lipophilic	Solid solution
Progesterone	Lipophilic	Molecular dispersion in vitreous medium
Hydrocortisone	Hydrophilic	Crystalline dispersion

Table 2
MICROSPHERES OF POLY(D,L-LACTIC ACID), RATE OF INCORPORATION OF DIFFERENT ACTIVE INGREDIENTS[57]

Active ingredient	Initial drug concentration[a] (%)	Incorporation rate[b] (%)
Lomustine	23	91
Progesterone	23	92
Hydrocortisone	23	55

[a] Percentage by weight of active ingredient/total weight of materials (polymer + active ingredient).
[b] Measured by HPLC.

where: dM_t/dt = rate of release in steady state conditions at time, t; l = membrane thickness; A = area of galenic form; D · K = membrane permeability; ΔC = difference between internal and external drug concentrations (Co in sink conditions).

The quantity of active ingredient released varies directly with time (Figure 20). The rate of release is independent of time and is only affected by geometric factors (pharmaceutical form) and physicochemical factors (drug, polymer). The kinetics is of zero order.

For a microcapsule bounded by a microporous polymeric film, the rate of release[1] is given by Equation 2:

$$\frac{dM_t}{dt} = \frac{A \cdot D_w \cdot \epsilon}{\tau} \cdot \frac{\Delta C_w}{l} \tag{2}$$

where: dM_t/dt = rate of release of active ingredient; A = area of galenic form; D_w = diffusion coefficient of active ingredient in the liquid phase (water) impregnating the pores; ΔC_w = concentration gradient of active ingredient in the liquid phase impregnating the pores; ε = membrane porosity; τ = tortuousness of the diffusion path through the membrane; l = membrane thickness.

The kinetics of release is of zero order as in the foregoing case, but with a number of differences. These include the introduction of porosity and tortuosity factors, solubility of the drug referring not to its polymeric phase but to the dissolution medium impregnating the pores, and the dropping of the coefficient K in Equation 2.

For a matrix form (microsphere) in which release occurs through the water-saturated

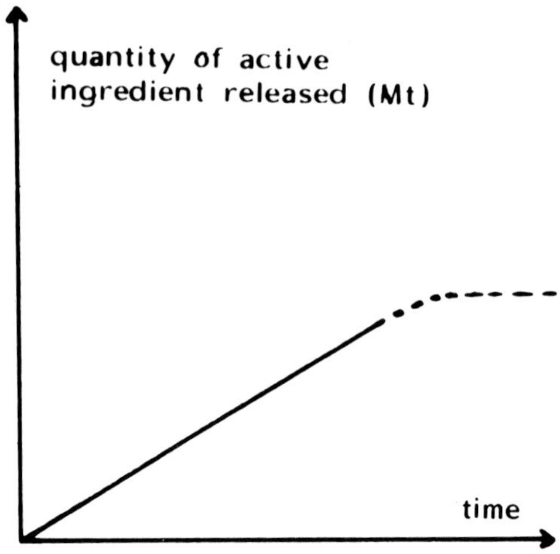

FIGURE 20. Microcapsule, profile or release kinetics.[57]

pores of the matrix, the rate of release is theoretically given by Equation 3, directly derived from Higuchi's model:[23]

$$\frac{dMt}{dt} = A \cdot {}^1/_2 \left\{ \frac{D_w \cdot \epsilon}{\tau} \cdot C_{sw} \, (2Co - \epsilon C_{sw}) \right\}^{1/2} \cdot t^{-1/2} \qquad (3)$$

where: dM_t/dt = rate of release of active ingredient; A = area of galenic form; D_w = diffusion coefficient of active ingredient in water; Co = initial concentration of active ingredient in matrix form; C_{sw} = solubility of active ingredient in water; ε = porosity; τ = tortuosity.

The quantity of active ingredient released is proportional to the square root of time (Figure 21).

To achieve satisfactory results, every in vitro investigation must be conducted using a protocol that satisfies a number of conditions:

1. Agitation of the pharmaceutical form must take place in laminar flow conditions.
2. The pharmaceutical form must not be eroded during handling.
3. After complete release, the total drug concentration in the external medium must be less than 10% of the saturation concentration (sink conditions).

As an illustration, Figure 22 represents the kinetics of release of lomustine obtained in our laboratory from microspheres of poly(D,L-lactic acid). The quantities of lomustine remaining in the particles were determined by the Griess reaction[38] and plotted as a function of time. The kinetic analysis was performed in the same way due to the rapid degradation of lomustine in aqueous medium at pH 7.2.[46] On the whole, the decrease profile of lomustine concentrations in the microspheres appears to show an exponential distribution. However, the interpretation of the kinetic results with this particular molecule remains difficult due to its instability. The diffusion of water and the effect of temperature (37°C) are two factors added to the simple diffusion of lomustine in the external medium.[5]

To identify the mathematical distribution describing the release of a lipophilic molecule from microspheres of poly(D,L-lactic acid), a more thorough study was conducted with progesterone.[7] The method described by Benita[2] is based on the principle of nonlinear regression,[8] which serves to pass as close as possible to the experimental points

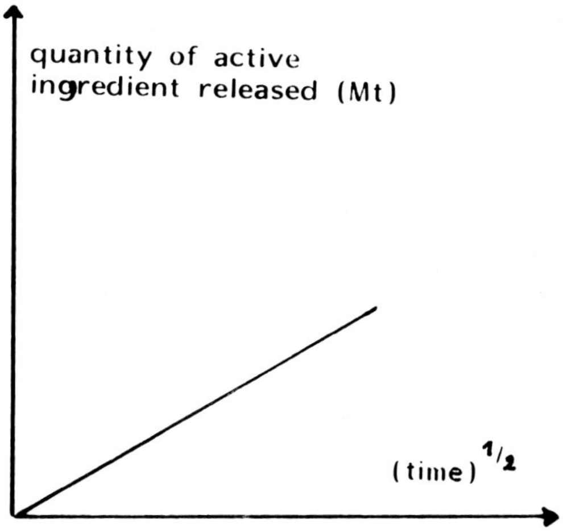

FIGURE 21. Microsphere, profile of release kinetics.[57]

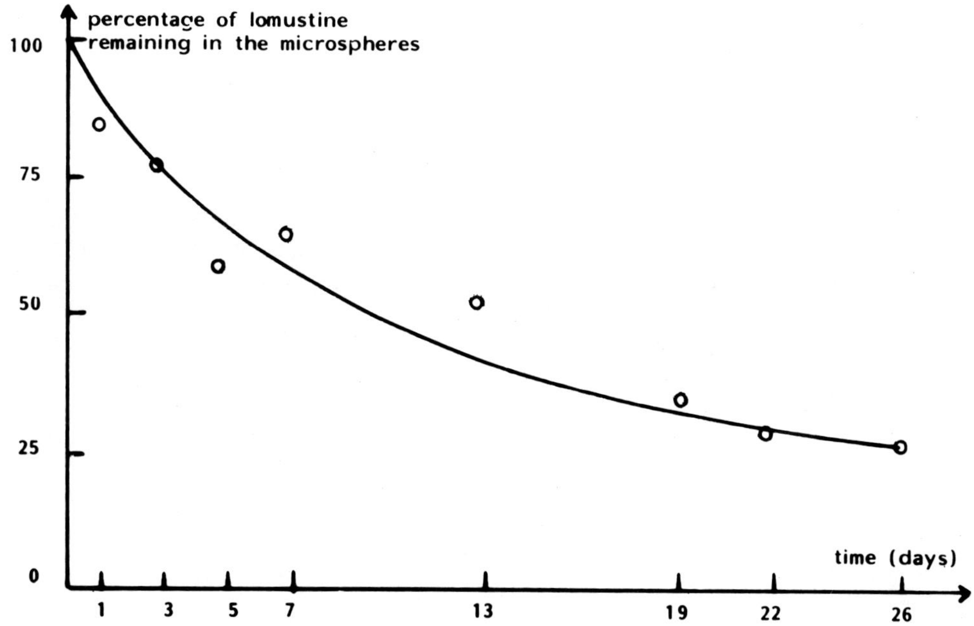

FIGURE 22. Microspheres of poly(D,L-lactic acid), in vitro release of lomustine expressed as a percentage remaining in the particles.[5]

and theoretical kinetic curves which obey different laws, the first order law, and the law derived from Higuchi's model. The χ^2 test is then applied to estimate the deviation between the experimental curve and the theoretical curves plotted previously. The resulting values of χ^2 concerning the first order equation:

$$M_t' = w_0(1 - e^{-k.t}) \tag{4}$$

and concerning the distribution deriving from Higuchi's model:

$$M_t = kt^{1/2} \tag{5}$$

are compared to determine which of the two models best fits the experimental data observed. A value of χ^2 equal to 0 implies that the kinetic data observed and the kinetic data anticipated from one of the two models are identical. On the whole, the lower value of χ^2 obtained in the comparison helps to assign the correct mathematical release model to the experimental data.

Figure 23 shows the experimental release kinetics of progesterone from microspheres slightly loaded with active ingredient (24.4%), as well as the two curves predicted from the first order and Higuchi distributions. Neither of the two kinetic models fits the experimental points. This result is confirmed by the values of χ^2 of 10.1 and 33.5 obtained respectively for the first order and Higuchi equations. This can be explained by the earlier investigations conducted by the scanning electron microscope and by DTA (Figures 11 and 16). These two investigations helped to show that progesterone is intimately dispersed in the polymer in the molecular state. This is also confirmed in a study by the transmission electron microscope (Figure 24). No pores can be detected even at high magnification, and the system is perfectly homogeneous. The release kinetics of progesterone from microspheres of poly(D,L-lactic acid) must therefore be comparable to the kinetics of desorption of an active ingredient dissolved in a spherical matrix. By using certain approximations, the initial part of the drug release curve in this physical state is described by the equation:[1]

$$\frac{M_t'}{w_0} = 6\left(\frac{Dt}{r^2\pi}\right)^{1/2} - 3\left(\frac{Dt}{r^2}\right) \tag{6}$$

where: M'_t = quantity of drug released in time, t; w_0 = initial quantity of drug in the microspheres; D = diffusion coefficient of drug in the polymer; r = radius of the microsphere. This equation is valid for a ratio $(M'_t/w_0) < 0.4$. The final part of the release curve is given by the expression:

$$\frac{M_t'}{w_0} = 1 - \frac{6}{\pi^2} e^{-\frac{\pi^2 Dt}{r^2}} \tag{7}$$

In fact, this equation is a first order equation.

Consequently, by taking account of these equations, the total release of progesterone from microspheres of poly(D,L-lactic acid) should follow neither a first order distribution nor a Higuchi distribution, but rather a composite distribution derived from both models. In fact, the analysis of the different portions of the experimental release curve according to Equations 6 and 7 shows that the initial quantity of drug released does not vary linearly with the square root of time. On the other hand, the final release of progesterone from the microspheres follows a first order distribution which agrees with Equation 7 (Figure 25).

The results obtained partly disagree with the theory of Baker and Lonsdale.[1] This occurs possibly because of the physical state of the encapsulated active ingredient. According to Baker's theory, the active ingredient is assumed to be dissolved in the matrix and hence in the state of a solid solution in the polymer. As shown by DTA analyses, progesterone is in a different state, that of a molecular dispersion in a vitreous medium. This could cause changes in the level of its release kinetics.

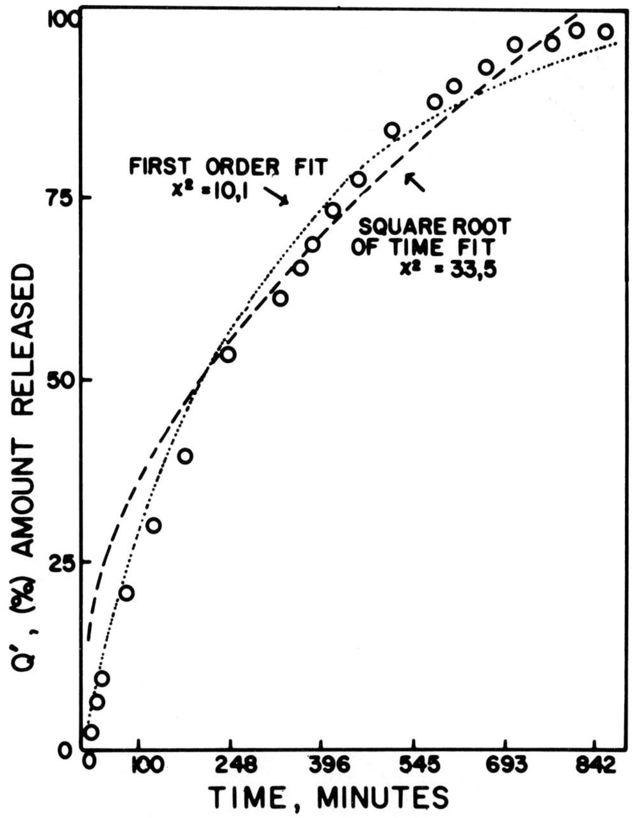

FIGURE 23. Microspheres of poly(D,L-lactic acid), in vitro release of progesterone from low-drug loaded (24%) microspheres.[7]

However, microspheres highly loaded with progesterone (68.3%) display a release profile corresponding to a first order kinetics (Figure 26). This is confirmed by the values of χ^2 of 0.99 and 67.00 respectively for the first order and Higuchi models. This observed kinetic profile tends to show that the release of progesterone from these microspheres occurs by dissolution rather than diffusion across a homogeneous matrix. These conclusions tend to support the observation made under the electron microscope (Figure 13).

It is interesting to note that the increase in the drug content, progesterone in this case, alters the structure of the microspheres and leads to different mathematical kinetic distributions. This entire study clearly reveals the influence of the physical state of the encapsulated active ingredient, namely of the cytostatic agent in the case of chemoembolization, on the modality of release.

G. Storage of Microspheres

In the specific case of microspheres of poly(D,L-lactic acid), the physical state of the encapsulated drug directly affects its stability within the microparticulate system (Table 3). If the active ingredient is dispersed in the molecular state, corresponding to a molecular dispersion in a vitreous medium or solid solution, interactions with the macromolecular chains are favored and degradation is normally more rapid (lomustine, progesterone). If the drug is dispersed in crystalline form, interactions with the polymer are diminished and stability is generally greater (hydrocortisone).[5]

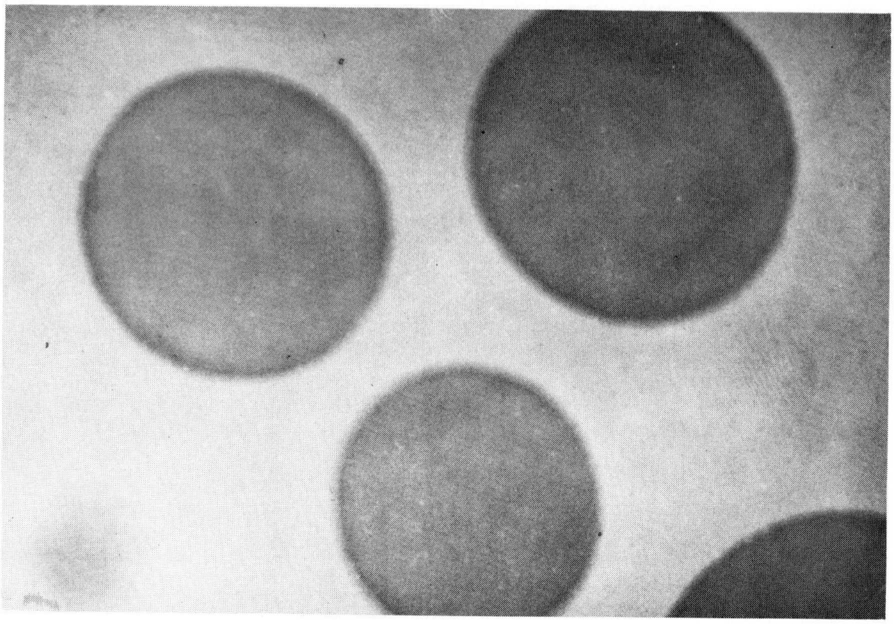

FIGURE 24. Microspheres of poly(D,L-lactic acid) containing 23% progesterone, microsections observed by the transmission electron microscope. Magnification × 9000.[5]

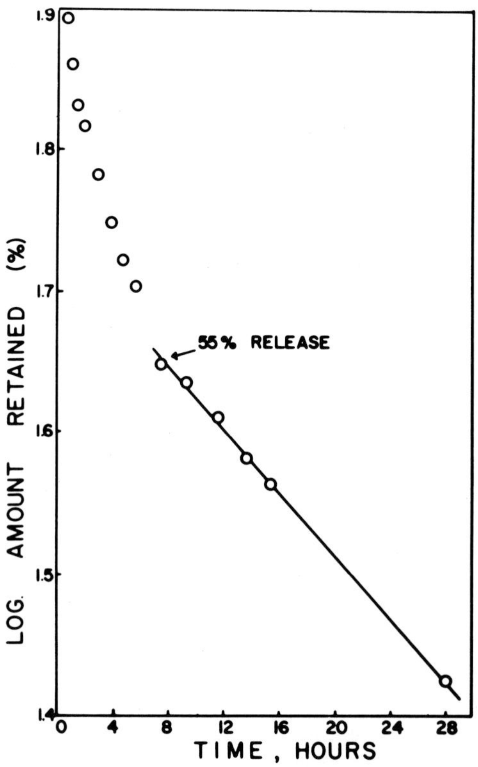

FIGURE 25. Microspheres of poly(D,L-lactic acid), in vitro release of progesterone from low-drug loaded (24%) microspheres.[7]

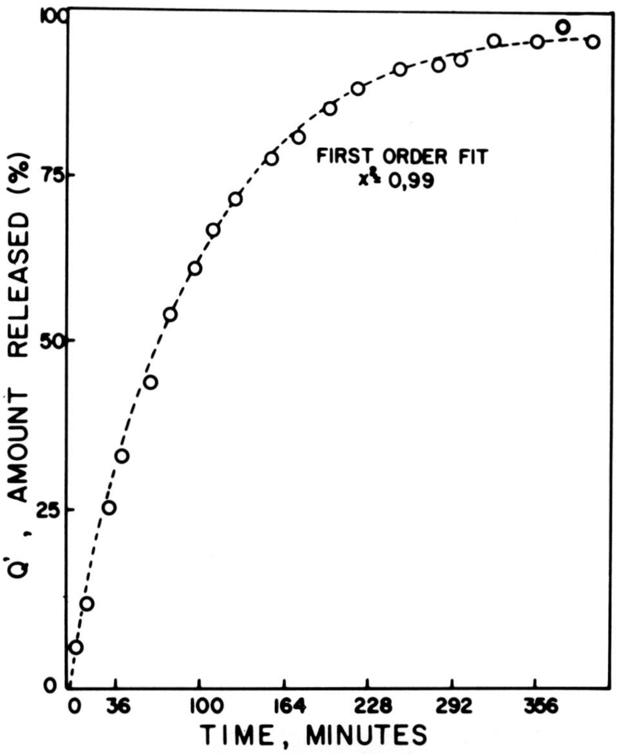

FIGURE 26. Microspheres of poly(D,L-lactic acid), in vitro release of progesterone from high-drug loaded (68%) microspheres.[7]

Table 3
MICROSPHERES OF POLY(D,L-LACTIC ACID), INFLUENCE OF PHYSICAL STATE OF ENCAPSULATED DRUG ON STORAGE STABILITY[5]

Active ingredient	Physical state of encapsulated drug	Storage stability[a]
Lomustine	Solid solution	Poor
Progesterone	Molecular dispersion in vitreous medium	Poor
Hydrocortisone	Crystalline dispersion	Good

[a] Storage at 25°C, in darkness, in a desiccator.

IV. ANIMAL AND CLINICAL INVESTIGATIONS OF MICROPARTICLES FOR EMBOLIZATION AND CHEMOEMBOLIZATION

A. Study in the Animal

1. Embolization

The microcapsules and microspheres tested were prepared in the laboratory from:

carnauba wax, cetyl alcohol, poly(D,L-lactic acid), poly(β-hydroxybutyric acid), 50/50 copolymer (lactic acid/glycolic acid), and serum/albumin. They were analyzed in the dog and rat to determine the following:

- The harmlessness of the support materials
- Optimal particle size
- Occlusive properties
- Further developments in the organism (biodegradability)[6,39,40]

The operating procedure is based on selective (renal artery) and hyperselective (polar artery) embolization of the kidney of the animal (Figure 27).

a. General Tolerance

No animal died from the after effects of embolization or lost weight significantly, between the day of embolization and the date of sacrifice, regardless of the material employed.[39,40]

b. Location of Microparticles in the Arterial System

The microparticle location was determined by arterial opacification and the use of microspheres that were made radio-opaque by the encapsulation of a contrast material. In immediate postembolization, the particle distribution in the arterial system depends according to particle size. The larger the microsphere size, the more proximal their location. However, due to the increasing blood pressure, the elasticity of the vascular walls, and a reactional vasoplegia subsequent to embolization, the microemboli migrate in the arterial system as long as the inflammatory granuloma with foreign bodies has not formed (see Section IV.A.1.c). By taking account of the diameter of the arteries and of the migratory phenomenon, a particle size range between 200 and 400 μm appears to be optimal for embolization. Microspheres that are too small migrate too far and go beyond the target. If too large, they cause a proximal occlusion leading to ischemia of the entire organ which is no longer related to the pathological target area.[39,40]

c. Occlusive Properties

The effectiveness of embolization was determined by an anatomo-pathological analysis of the embolized tissues[39,40] and also by a biochemical analysis in the specific case of microcapsules of cross-linked serum/albumin.[6] With microspheres of carnauba wax and cetyl alcohol, the anatomo-pathological lesions in the kidney vary with time after embolization. A recent infarctus and a fibrocruoric clot appear in the first week. They are succeeded in the following weeks by an older infarctus and an inflammatory reaction with foreign matter. This foreign matter reaction appears earlier or later depending on the nature of the support material. It is established on day 8 with carnauba wax (Figure 28), and earlier with cetyl alcohol. This reaction is less intense with poly (D,L-lactic acid).

It should be noted that this local reaction is not accompanied by an immunorejection phenomenon. It also serves to stop the particle definitively in its migration in the vascular bed. It should also be noted that anatomo-pathological investigations under way have revealed fibrous endarteritis and periarteritis lesions after embolization with microspheres and microcapsules of different types. However, it is difficult to state whether these lesions result from the action of the microparticles or from the excipient employed for the injection.

Lesions of the same type have been observed in the mononephrectomized rat after embolization of the kidney with microcapsules of cross-linked serum/albumin.[6] More-

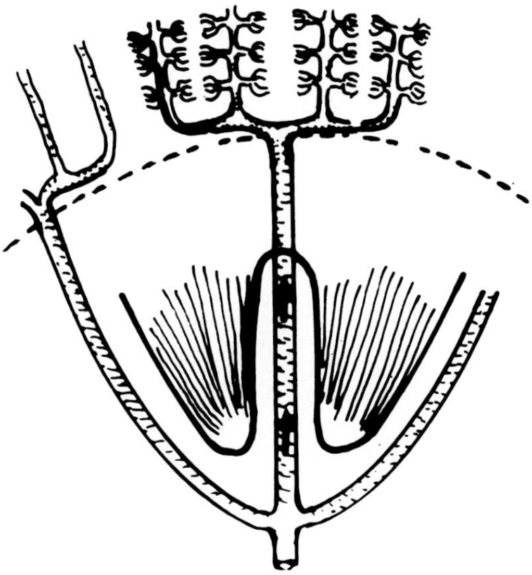

FIGURE 27. Microcapsules and microspheres for chemoembolization, arterial system of the rat kidney.[6]

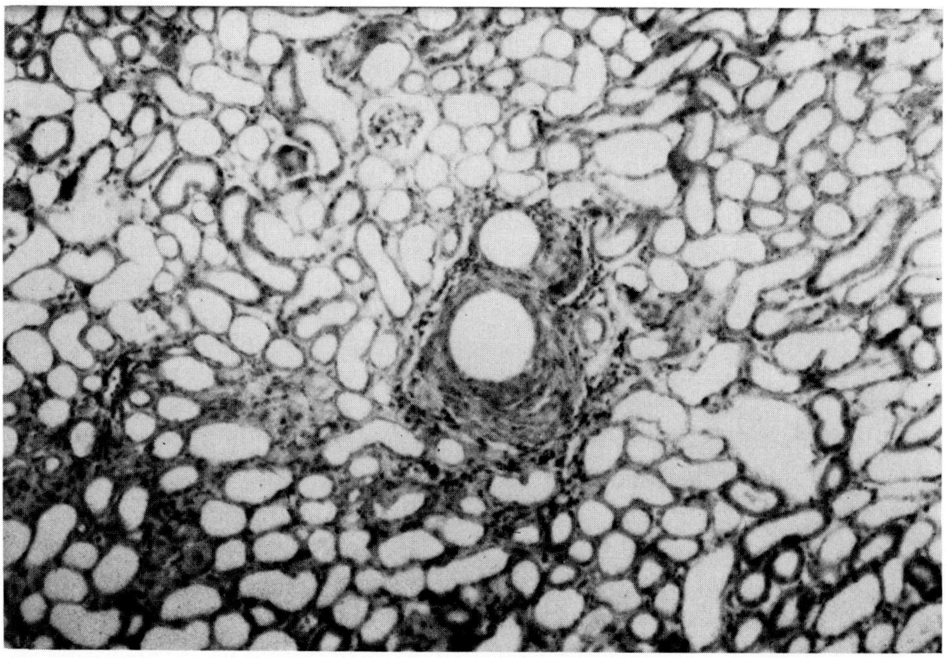

FIGURE 28. Carnauba wax microspheres, inflammatory granuloma appearing around microparticles injected into the renal artery of the dog.[57]

over, the quality of embolization was assessed by observing the concentration of plasma creatinin, which related to the renal function. The results are summarized in Table 4.

Table 4
MICROCAPSULES OF CROSS-LINKED SERUM/ ALBUMIN, QUALITY OF EMBOLIZATION AS A FUNCTION OF INJECTED MICROCAPSULE DOSE

Volume injected (mℓ)	Creatinin concentration (μmol/ℓ)	Renal insufficiency
0.75 to 1.5	120	Acute but reversible
1.5 to 2	120 to 300	Chronic, irreversible
Over 2	450 to 600	Acute, irreversible

Note: Normal creatinin concentration: 30 to 40 μmol/ℓ.

The quality of embolization depends on the volume injected and hence on the quantity of microcapsules injected (concentration of 3 cg of microcapsules per 10 mℓ of physiological serum/contrast material). For a volume between 0.75 and 1.5 mℓ, the creatinin rate is three to four times higher than normal. After 1 week, creatinemia returns to its physiological level. Renal insufficiency is acute but reversible and embolization is partial. For a volume between 1.5 and 2 mℓ, the values of creatinemia are multiplied by a factor ranging from 3 to 10 and there is no return to the physiological levels. Renal insufficiency is chronic but irreversible and embolization tends to be complete. For an injected volume exceeding 2 mℓ, the creatinemia rates are multiplied by 15. Renal insufficiency is acute and irreversible, and embolization is complete and causes the death of the mononephrectomized animals in 3 days.

d. Biodegradability

Biodegradability depends on the nature of the material used. Microspheres of carnauba wax are absolutely not attacked in the vascular bed of the dog, even after *in situ* residence of 1 year.[39] As for microspheres of cetyl alcohol, fragmentation occurs from day 5.[63] Given the chemical composition of the materials, this is a case of disintegration rather than biodegradability.

Investigations under way show that poly(D,L-lactic acid) and poly(β-hydroxybutyric acid) are not attacked within 15 days following embolization of the dog's kidney. On the other hand, microspheres of 50/50 (lactic acid/glycolic acid) copolymer reveal degradation on the 7th day after embolization. The most promising results are nonetheless those obtained with microcapsules of cross-linked serum/albumin. Depending on the degree of cross-linkage, their biodegradability can be adjusted. Two types of investigations (renal embolization) have been conducted into these particles. In the rat, microcapsules with a degree of cross-linkage corresponding to 1 g of acyl chloride per gram of serum/albumin are resorbed within 15 to 28 days.[6] In the dog, microcapsules with a degree of cross-linkage corresponding to 0.25 g acyl chloride per gram of serum/albumin disappear from the vascular bed in less than 7 days (investigation under way).

It is important to determine the biodegradability of microparticles inasmuch as it serves to select materials that can be used for chemoembolization. A nonbiodegradable material causes permanent obliteration of the vascular bed. This is the objective of embolization. A biodegradable material is resorbed and releases its medicinal content, thus allowing continued additions of microparticles at the target tissue. This is the objective of chemoembolization.

2. Chemoembolization

The study of chemoembolization is conducted by foreign teams on the healthy animal and on the tumoral animal.

a. *Healthy Animal*

The first investigations were published by Lindell et al.[37] These authors examined the pharmacokinetics of inulin released from microspheres of cross-linked starch (20 μm in diameter), which were placed in the hepatic artery of the rat.[37] The plasma concentration of inulin in the peripheral blood decreases significantly in the few minutes following the injection of the microspheres and is far more constant than that resulting from a simple intra-arterial injection of the drug. This is confirmed by the percentage of inulin excreted as a function of time. If inulin is injected alone, 60% is eliminated in 35 min. In the same interval, only 20% is excreted if the inulin is microencapsulated. These authors examined the toxicity of 5-FU administered to the healthy rat alone or encapsulated (hepatic embolization).[37] The toxicity was estimated by a count of the leukocytes and by determining the mortality. With microencapsulated 5-Fu, 75% of the rats survived, as compared with 25% with 5-FU alone.

Kato's team evaluated the obliterating and chemotherapeutic effects on the dog of ethylcellulose microcapsules loaded with mitomycin C. These microcapsules have a mean diameter of 224 μm. The kidney was embolized in one case and the pelvic organs in the other (rectum, prostate, bladder, lymphatic ganglia). With renal embolization, the obliterating effects were determined by anatomo-pathology and angiography.[27,48] The histological analysis showed that the microcapsules of mitomycin C cause necrosis lesions of the entire cortex and of two thirds of the medullary. The lesions observed are significantly larger than those caused by neutral microcapsules injected with or without mitomycin C solution. In a preliminary study conducted ex vivo, angiography showed a decrease in distal vascularization of the kidney.[26] This was confirmed subsequently. Microcapsules of mitomycin C cause complete obliteration of the arcuate arteries. The proximal arteries are not embolized. One interesting conclusion provided by these investigations concerns the influence of encapsulated mitomycin C on the quality of obliteration. The presence of the drug in the embolus causes larger necrosing ischemic lesions than those due to the embolism alone. Some investigators[48] have postulated a biochemical reaction of mitomycin C with the endothelial cells of the vascular wall, as well as direct cytological action of the drug on the surrounding renal tissue.

The chemotherapeutic effects of mitomycin C microcapsules were followed up by taking kidney and serum samples and by estimating the cytostatic power of organ and serum brei on cultures of *Escherichia coli.* The results obtained show that the kidney brei embolized with chemoemboli retain cytostatic power 6 hr after the injection, whereas, in the same interval, mitomycin C cannot be detected in the kidney infused with the drug alone. Plasma concentrations of the active ingredient in the peripheral blood are sharply reduced if the mitomycin C is encapsulated. The quantity of antitumor agent released in the blood compartment from microcapsules only accounts for 43% of the quantity present after intra-arterial injection of the same dose of mitomycin C alone. The activity of the encapsulated drug found in the renal tissue 6 hr after the start of embolization proves that ethylcellulose microcapsules protect the active ingredient from the powerful enzyme inactivation exerted by the kidney on this molecule. Moreover, the reduced bioavailability of mitomycin C for the blood compartment implies a possible reduction in systemic toxicity to tissues such as the hematopoietic tissues, for example.

Kato et al. also similarly evaluated the chemotherapeutic effects of mitomycin C microcapsules on the pelvic organs of the dog, the rectum, prostate, bladder, and lymphatic ganglia.[49] The results are closely comparable with those obtained after renal embolization: presence of cytostatic activity in embolized organ brei and a plasma concentration of mitomycin C of 45% of the level present after intra-arterial injection of the drug alone.

b. Animal with a Tumor

Investigations conducted on the animal with an experimental catheterizable tumor are extremely restricted due to the technical problems raised by these experiments. However, a recent study helped to demonstrate the effectiveness of albumin microspheres loaded with mitomycin C on the treatment of tumors.[20] The animal selected was a rabbit in which a VX-2 experimental carcinoma was implanted in the hind paw. The growth of the tumor, the mortality, and the plasma and tissue concentrations of mitomycin C in the tumor were assessed. After the injection of mitomycin C microspheres, the growth of the tumor was stopped and the tumor disappeared in 3 weeks. The mortality rate was 100% in the rabbit treated with neutral microspheres or with nonencapsulated mitomycin C. It was only 50% after treatment with the chemoemboli.

The drug plasma concentrations in the peripheral blood displayed different profiles according to whether the mitomycin C was encapsulated or not. Its incorporation in microspheres served to reduce the blood concentrations considerably, while prolonging them in time. Finally, mitomycin C concentrations in the tumor were sharply increased; this was observed over an interval of 8 hr.

B. Clinical Investigation

1. Embolization

Merland et al. employed neutral carnauba wax microspheres in therapeutic angiography.[42,43] Patients, (18), with a wide variety of ailments (tumors, vascular malformations) were treated.[43] Follow-up tests of vascular occlusion dealt with several criteria:

- Control arteriography
- Resulting clinical improvement
- Operational observations
- Histological observations

In all cases, the arteriography, with immediate improvement, revealed an arteriole lesion occlusion. Surgery ensued in six cases. Due to the quality of obliteration, the operative risk was significantly reduced, facilitating surgery. In other inoperable cases, embolization by microspheres was effective in obliterating intratumor vessels and the clinical symptomatology was improved. When the vessels to be obliterated were too large (especially arteriovenous shunts), the microspheres employed failed to achieve positive therapeutic results due to their size.

2. Chemoembolization

To the best of our knowledge, Kato and his team are the only ones to have published clinical results on the preoperative and curative treatment of cancer by chemoembolization. The tumors treated are of different origins and the procedure differed according to the type of tumor: renal carcinoma or primary and secondary carcinomas of the liver, bones, and pelvic organs. However, whatever the nature of the tumor, the chemoembolus always consisted of an ethylcellulose microcapsule containing mitomycin C in a concentration of 80%.[28,50]

a. Renal Carcinoma

This tumor is a specific case because it displays many arteriovenous fistulas and hypervascularity.[28] Hence the mitomycin C microcapsules are injected simultaneously with pieces of resorbable gelatin sponge (Gelfoam), through the femoral pathway. According to Kato, this technique serves to obliterate all the feeder vessels of the tumor.

To assess the effectiveness of the procedure described above, the patients were divided into two groups. The first group received a suspension of microcapsules of mitomycin C and Gelfoam in physiological serum. The second group of patients received a suspension of Gelfoam in a saline solution of mitomycin C. The doses of encapsulated and free mitomycin C were identical in both procedures.

The effectiveness of treatment was assessed in various ways. The decrease in tumor volume was determined by tomography and echography. Embolization effectiveness was observed by angiography. Anatomo-pathological analysis served to evaluate tissue necrosis. The plasma concentrations of mitomycin C and the blood concentrations of the blood elements (hematotoxicity) were determined in the peripheral blood.

As an illustration, a number of clinical results are presented, based on 43 patients, of whom 33 were treated by chemoembolization.[28] Of the patients treated by chemoembolization, 68% showed a significant decrease in the volume of the tumor. None of the patients treated simultaneously by embolization and chemoinfusion displayed any comparable decrease in tumor mass. Tumor necrosis was only perceptible 4 to 6 weeks after the operation.

Chemoembolization alone led to persistent vascular occlusion (33% of the patients). Embolization by Gelfoam, even if supplemented by mitomycin C infusion, led to repermeation of the embolized vessels and/or the development of collateral circulation. Anatomo-pathology showed that a considerable necrosis developed around the vessels containing fragments of mitomycin C microcapsules.

The decrease in tumor mass associated with virtually complete vascular occlusion facilitates the surgery of highly invasive tumors. Hematotoxicity was significantly decreased in the patients treated by chemoembolization (21% instead of 50%). In the case of chemoembolization, the plasma concentrations of antitumor agents were sharply decreased in comparison with those obtained after embolization and intra-arterial infusion of mitomycin C. These concentrations amounted to 39% of the plasma concentrations of mitomycin C infused in free form. This shows that, among other advantages, chemoembolization helps to avoid a harmful systemic impregnation that is responsible for toxic secondary effects.

b. Primary and Secondary Carcinomas of the Liver, Bones, and Pelvic Organs[29, 30]

In these pathological cases, chemoembolization is practiced by means of mitomycin C microcapsules in the absence of any other embolus. Embolization is achieved after a single or multiple injection of the microcapsules and the procedure depends on the size of the tumor and the patient's overall condition.

The effectiveness of the treatment is reflected by the different results:

1. Decrease in tumor volume (tomography, echography)
2. Obliteration (angiography)
3. Necrosis (anatomo-pathology)
4. Survival rate

Systemic toxicity was also analyzed directly by hematological examinations and indirectly by determining the plasma concentrations of mitomycin C in the peripheral blood.

Out of 33 patients treated by chemoembolization, a significant tumor reduction was achieved in 61% of the cases. Pain was relieved in 80% of the cases.[30] Hematotoxicity only occurred in 23% of the cases.[30]

A second clinical study served to evaluate the survival rate over a period ranging from 5 to 21 months.[29] Out of 20 patients, two thirds of whom failed to respond positively to conventional treatment, 12 patients survived.

V. CONCLUSIONS

The rapid clearance of anti-cancer drugs from tumor tissues, together with their secondary effects, present major problems in anti-cancer chemotherapy. A possible approach designed to overcome these problems is the selective administration of a pharmaceutical form that is inherently active or that allows controlled release of the drug directly in the target pathological tissues. Microcapsules and microspheres for embolization and chemoembolization offer one of the examples that illustrate this approach.

This chapter has dealt with the following subjects:

- Basic concepts concerning the therapeutic techniques of embolization and chemoembolization
- Technological aspects of microcapsules and microspheres for embolization and chemoembolization
- Their pharmacological effect in the animal and their clinical effectiveness

A considerable body of work still needs to be done, especially in the field of chemoembolization. According to certain oncologists, it is important to use microparticles loaded with cytostatic agents capable of biodegrading within a very short interval, so as to avoid damaging the vascular endothelium irreversibly. In no circumstances should the vascular access to the tumor be obliterated, because it allows the repeated injection of drug-loaded microparticles. Research trends in the near future will aim at a more thorough investigation and the development of cross-linked serum-albumin microcapsules, which are biodegradable in a short interval and capable of conveying a large number of water-soluble antimitotic substances.

ACKNOWLEDGMENTS

Part of the work cited in this chapter formed or forms the subject of theses by M. C. Bissery, J. P. Benôit, R. Fickat, P. Hoffmann, D. Terracol, P. Tilleul, and O. Zouaï.

REFERENCES

1. Baker, R. W. and Lonsdale, H. K., Controlled release: mechanisms and rates, in *Controlled Release of Biologically Active Agents,* Tanquary, A. C. and Lacy, R. E., Eds., Plenum Press, New York, 1974, 15.
2. Benita, S., Kinetic model identification of drug release from microcapsules using the differential rate method or the non-linear regression search procedure, in Proc. 5th Int. Symp. on Microencapsulation, Montreal, May 1983.
3. Benita, S., Benoit, J. P., Puisieux, F., and Thies, C., Characterization of drug-loaded poly (d,ℓ-lactide) microspheres, *J. Pharm. Sci.,* 73(12), 1721, 1984.
4. Benoit, J. P., Puisieux, F., Thies, C., and Benita, S., Characterization of drug-loaded poly(d,ℓ-lactide) microspheres, in 3rd Int. Conf. on Pharmaceutical Technology, Vol. 3, Paris, 1983, 240.
5. Benoit, J. P., Préparation et caractérisation de microsphères biodégradables pour chimioembolisation. Thèse Doctorat d'Etat, Paris IX, 1983.
6. Benoit, J. P., Drouin, J. Y., Puisieux, F., Brunelle, F., Dubois, M., and Beaufils, M., Selective embolization of the renal artery of the rat by cross-linked serum/albumin microcapsules, in *Microspheres and Drug Therapy,* Elsevier, Amsterdam, 1983.
7. Benoit, J. P., Benita, S., Puisieux, F., and Thies, C., Stability and release kinetic analysis of drugs incorporated within microspheres, in *Microspheres and Drug Therapy,* Elsevier, Amsterdam, 1983, 91.

8. Bevington, P., *Data Reduction and Error Analysis for the Physical Sciences,* McGraw Hill, New York, 1969.
9. Bissery, M. C., Valeriote, F., Puisieux, F., and Thies, C., In vitro and in vivo evaluation of CCNU-loaded microspheres prepared from poly(d,ℓ-lactide) and poly(β-hydroxybutyrate), in *Microspheres and Drug Therapy,* Elsevier, Amsterdam, 1983, 217.
10. Bonnel, D., Helenon, C., and Bigot, J. M., L'embolisation dans le traitement des hémoptysies, *Vie. Med.,* 13, 895, 1982.
11. Brooks, B., Treatment of Traumatic arteriovenous fistulas, *Sth. Med. J.,* 23, 100, 1930.
12. Chabner, B. A. and Myers, C. E., Clinical pharmacology of cancer chemotherapy, in *Cancer, Principles and Practice of Oncology,* Devita, V. T., Hellman, S., and Rosenberg, S. A., Eds., J. B. Lippincott, 1982, 156.
13. Champetier, G. and Monnerie, L., *Introduction à la Chimie Macromoléculaire,* Masson, Paris 1969.
14. Chang, T. M. S., *Artificial Cells,* Charles C Thomas, Springfield, Ill., 1972.
15. Chiou, W. L. and Riegelman, S., Pharmaceutical applications of solid dispersion systems, *J. Pharm. Sci.,* 60(9), 1281, 1971.
16. Couvreur, P., Mise au point d'un nouveau vecteur de médicament, Thèse d'Agrégation, Université Catholique de Louvain, Louvain, Belgium, 1983.
17. Crooke, S. T. and Bradner, W. T., Mitomycin C: a review, *Cancer Treat. Rev.,* 3, 121, 1976.
18. Curet, P., Roche, A., Weingarten, A., and Doyon, D., L'embolisation artérielle et veineuse thérapeutique, *Vie Med.,* 8, 569, 1979.
19. Djindjian, R., Houdart, R., Cophignon, J., Hurth, M., and Comoy, J., Premiers essais d'embolisation par voie fémorale de fragments de muscle dans un cas d'angiome médullaire et dans un cas d'angiome alimenté par la carotide externe, *Rev. Neurol.,* 125, 119, 1971.
20. Fujimoto, S., Endoh, F., Kitsukawa, Y., Okui, K., Morimoto, Y., Sugibayashi, K., Miyakawa, A., and Suzuki, H., Continued in vitro and in vivo release of an antitumor drug from albumin microspheres, *Experimentia,* 39, 913, 1983.
21. Gyves, J. W., Ensminger, W. D., Van Harken, D., Niederhuber, J., Stetson, P., and Walker, S., Improved regional selectivity of hepatic arterial mitomycin by starch microspheres, *Clin. Pharmacol. Ther.,* 34(2), 259, 1983.
22. Heilmann, K., *Therapeutic Systems,* Georg Thieme Publ., Stuttgart, 1978.
23. Higuchi, T., Mechanism of sustained-action medication, *J. Pharm. Sci.,* 52, 1145, 1963.
24. Hoffmann, P., Microemboles pour angiographie thérapeutique: mise au point et étude de microcapsules d'éthylcellulose contenant une substance cytostatique, le 5-fluorouracile, Thèse Doctorat de 3ème Cycle, Paris XI, 1979.
25. Kato, T. and Nemoto, R., Microencapsulation of mitomycin C for intra-arterial infusion chemotherapy, *Proc. Jpn. Acad.,* 548(71), 413, 1978.
26. Kato, T., Nemoto, R., and Nishimoto, T., Ex vivo intra-arterial infusion of microencapsulated mitomycin C into dog kidney, *Tohoku J. Exp. Med.,* 127, 99, 1979.
27. Kato, T., Nemoto, R., Mori, H., and Kumagai, I., Sustained release properties of microencapsulated mitomycin C with ethylcellulose infused into the renal artery of the dog kidney, *Cancer,* 46, 14, 1980.
28. Kato, T., Nemoto, R., Mori, H., Takahashi, M., and Tamakawa, Y., Transcatheter arterial chemoembolization of renal cell carcinoma with microencapsulated mitomycin C, *J. Urol.,* 125, 19, 1981.
29. Kato, T., Nemoto, R., Mori, H., Takahashi, M., and Harada, M., Arterial chemoembolization with mitomycin C microcapsules in the treatment of primary or secondary carcinoma of the kidney, liver, bone and intrapelvic organs, *Cancer,* 48, 674, 1981.
30. Kato, T., Nemoto, R., Mori, H. Takahashi, M., Tamakawa, Y., and Harada, M., Arterial chemoembolization with microencapsulated anticancer drug, *J. Am. Med. Assoc.,* 245, 1123, 1981.
31. Kerber, C. W., Bank, W. O., and Horton, J. A., Polyvinyl alcohol foam: prepackaged emboli for therapeutic embolization, *Am. J. Roentgenol.,* 130, 1193, 1978.
32. Kohn, K. W., Interstrand cross-linking of DNA by 1,3-bis(2-chloroethyl)-1-nitrosourea and other 1-(2-haloethyl)-1-nitrosoureas, *Cancer Res.,* 37, 1450, 1977.
33. Kunstlinger, F., Brunelle, F., Chaumont, P., and Doyon, D., Vascular occlusive agents, *Am. J. Radiol.,* 136, 151, 1981.
34. Lasjaunias, P. and Doyon, D., L'embolisation dans le territoire cervico-facial, *Vie Med.,* 13, 883, 1982.
35. Le Peq, J. B., Mécanisme d'action des substances antitumorales, in *Chimiothérapie Anticancéreuse,* Le Peq, J. B., Ed., Hermann, Paris, 1978, 23.
36. Lévy, M. C., Rambourg, P., and Lévy, J., Microencapsulation II: préparation de microcapsules à paroi constituée de polymères biodégradables, in Proc. 2nd Int. Conf. on Pharmaceutical Technology, Paris, Vol. 3, 1980, 15.
37. Lindell, B., Aronsen, K. F., Nosslin, B., and Rothman, U., Studies in pharmacokinetics and tolerance of substances temporarily retained in the liver by microsphere embolization, *Ann. Surg.,* 187(1), 95, 1978.

38. Loo, T. L. and Dion, R. L., Colorimetric method for the determination of 1,3-bis(2-chlorethyl)-1-nitrosourea, *J. Pharm. Sci.*, 54, 809, 1965.
39. Madoulé, P., Trampont, P., Doyon, D., Quillard, J., and Puisieux, F., Expérimentation chez le chien de microbilles utilisables en angiographie thérapeutique, *J. Radiol.*, 62(8—9), 457, 1981.
40. Madoulé, P., Trampont, P., Quillard, J., Doyon, D., Laval-Jeantet, M., Zouai, O., and Tilleul, P., Expérimentation de nouveaux materiaux d'embolisation, *Sci. Technol. Pharm.*, 11(10), 441, 1982.
41. Merland, J. J., Bories, J., Fredy, D., Thiebot, J., Tubiana, J. M., and Launay, M., L'embolisation ou thrombose thérapeutique endo-artérielle: une acquisition récente en radiologie vasculaire, *Rev. Med.*, 11—12, 683, 1976.
42. Merland, J. J., Bories, J., Simon, P., Puisieux, F., and Raziel, A., Une nouvelle étape en angiographie thérapeutique: les microembolisations, *Labo Pharma*, 281, 928, 1978.
43. Merland, J. J. and Riche, M. C., L'utilisation des microsphères dans le traitement par embolisation des tumeurs et des angiomes, *Suivi Thérapeutique, APHIF*, 45, 1981.
44. Merland, J. J. and Riche, M. C., Place de l'embolisation dans le territoire crânio-encéphalique et vertébro-médullaire, *Vie Med.*, 13, 871, 1982.
45. Miller, R. A., Brady, J. M., and Cutright, D. E., Degradation rates of oral resorbable implants (polylactates and polyglycolates): rate modification with changes in PLA/PGA copolymer ratios, *J. Biomed. Mater. Res.*, 11, 711, 1977.
46. Montgomery, J. A., Ruby, J., McCaleb, G. S., and Johnston, T. P., The modes of decomposition of 1,3-bis(2-chloroethyl)-1-nitrosourea and related compounds, *J. Med. Chem.*, 10, 668, 1967.
47. Montgomery, J. A., Johnston, T. P., and Fulmer Shealy, Y., Drugs for neoplastic diseases, in *Burger's Medicinal Chemistry*, Vol. 2, Wolff, M. E., Ed., John Wiley & Sons, New York, 1979, 595.
48. Nemoto, R., Kato, T., Mori, H., and Harada, M., Microencapsulated mitomycin C in transcatheter embolization of dog kidney, *Invest. Urol.*, 18, 69, 1980.
49. Nemoto, R. and Kato, T., Experimental intra-arterial infusion of microencapsulated mitomycin C into pelvic organs, *Br. J. Urol.*, 53, 225, 1981.
50. Nemoto, R., Kato, T., Iwaka, K., Mori, H., and Takahashi, M., Evaluation of therapeutic arterial embolization in renal cell carcinoma using microencapsulated mitomycin C, *Urology*, 17, 315, 1981.
51. Pharmacopée Francaise, Sterilisation par l'oxyde d'éthylène de matériel médico-chirurgical, IX ed., 1972, II-212-7.
52. Pharmacopée Francaise, Stérilisation par les rayonnements ionisants de mâteriel medico-chirurgical à usage unique et d'articles de pansement et de suture, IX ed., 1972, II-212-13.
53. Pharmacopée Francaise, Supplément à la IX ed., Monographie Microcapsules, J.O., 1978, 8233.
54. Puisieux, F. and Drouin, J. Y., Formes galeniques du futur, *Suivi Thérapeutique, APHIF*, 33, 1981.
55. Puisieux, F. and Raziel, A., Microspheres pour embolisation, *Sci. Technol. Pharm.*, 11(4), 149, 1982.
56. Puisieux, F. and Benita, S., Towards a new galenic dosage form, the liposome?, *Biomedicine*, 36, 4, 1982.
57. Puisieux, F. and Benoit, J. P., Systèmes micro- et nanoparticulaires, vecteurs de médicaments, Problèmes pharmacotechniques et analytiques, in *Voies Nouvelles de l'Evaluation Scientifique et Réglementaire du Médicament, 2ème Colloque INSERM/DPHM*, INSERM, 1983, 112, 81.
58. Raziel, A., Puisieux, F., Terracol, D., Hoffmann, P., Cave, G., Merland, J. J., and Bories, J., Wax microemboli tailored for therapeutic embolization, *Am. J. Roentgenol.*, 134, 404, 1980.
59. Reed, D. J., May, H. E., Boose, R. B., Gregory, K. M., and Beilstein, M. A., 2-chloroethanol formation as evidence for 2-chloroethyl alkylating intermediate during chemical degradation of 1-(2-chloroethyl)-3-cyclohexyl-1-nitrosourea and 1-(2-chlorethyl)-3-(trans-4-methylcyclohexyl)-1-nitrosourea, *Cancer Res.*, 35, 568, 1975.
60. Roche, A. and Brunelle, F., L'embolisation en pathologie digestive, *La Vie Médicale*, 13, 911, 1982.
61. Ross, G., Reul, B., Tillmann, W., and Liebenhoff, R., Vorrichtung zur Herstellung von perlformigen Granulat, Deutsches Patentamt, Patentschwift, 1918685, 1969.
62. Skiens, W. E., Burton, F. G., and Duncan, G. W., Biodegradable delivery systems, in *Biodegradables and Delivery Systems for Contraception*, Hafez, E. S. E., Ed., MTP Press limited, Lancaster, 1979, 3.
63. Terracol, D., Microemboles pour angiographie thérapeutique, Mise au point et étude de microcapsules de polyamide et de microbilles de cire contenant une substance cytostatique, le 5-fluorouracile, Thèse Doctorat de 3ème Cycle, Paris XI, 1979.
64. Thies, C., Microcapsules as drug delivery devices, *CRC Crit. Rev., Biomed. Eng.* 8(4), 335, 1982.
65. Tilleul, P., Microsphères d'alcool cétylique pour embolisation, etude et mise au point, incorporation de lomustine, premières applications, Thèse Doctorat de 3ème Cycle, Paris XI, 1983.
66. Vert, M., Chabot, F., Leray, J., and Christel, P., Stereoregular bioresorbable polyesters for orthopaedic surgery, *Makromol. Chem. Suppl.*, 5, 30, 1981.
67. Wise, D. L., Fellmann, T. D., Sanderson, J. E., and Wentworth, R. L., Lactic/glycolic acid polymers, in *Drug Carriers in Biology and Medicine*, Academic Press, New York 1979.

Chapter 6

COLLOIDAL PARTICLES AS RADIODIAGNOSTIC AGENTS

S. S. Davis, M. Frier, and L. Illum

TABLE OF CONTENTS

I. RADIODIAGNOSTIC COLLOIDS

A. Introduction

Colloidal particles in the form of microspheres, emulsions, suspensions, and liposomes can be labeled with gamma-emitting radionuclides and thereby find use as radiodiagnostic agents which are administered by a variety of routes. Most of the different applications can be categorized as "scanning" where the colloidal particles can be concentrated by a given normal tissue site but excluded or having a reduced uptake in an abnormal or pathological structure, or vice versa. Other radiodiagnostic applications include studies on kinetic processes such as uptake of particles by the cells of the reticuloendothelial system, blood and lymph flow measurements, and gastrointestinal transit. Visualization and quantification are achieved by placing the subject in front of a gamma camera or rectilinear scanner.[1] The radionuclides of choice for scintigraphic imaging are those that have a suitable energy for detection by the gamma camera (in the range 100 to 400 keV) and a half-life that is long enough to allow the objective of the study to be achieved but short enough to minimize the radiation dose. Table 1 gives details of some of the radionuclides that can be used in studies in man, as well as animal experiments where questions of radiation dose are less relevant.

Nonlabeled colloidal systems can be used as contrast agents in related diagnostic procedures such as computed tomography (CT) and NMR imaging. For example, the ideal contrast agent for improving the detection of small tumors in the liver using a CT scanner is one that absorbs X-rays efficiently, is well taken up by the liver, opacifies selectively either normal or pathological tissues, and is nontoxic.

B. Localization of Colloids

The basis of localization of colloids at tissue sites for scanning and diagnostic purposes can be considered in terms of three separate biological processes provided they are substrate nonspecific[2,3] (Table 2). They range in order from the simple physical placement of the colloid at a given site to a biological mechanism of localization. In some cases more than one localization mechanism may be operative, for example, accumulation of colloidal particles in lymph nodes can occur as a result of both entrapment (filtration) and phagocytosis.[4] Each localization mechanism will be discussed briefly.

1. Compartmental Localization

The simplest way of localizing a radiodiagnostic colloid is to provide retention of the agent in a tissue or body fluid space for a significant period of time. Thus, administration of a material in the gastrointestinal tract, the blood circulation, or the lymphatic system comes into this category.

2. Capillary Blockage

The blood capillaries range from about 7 to 10 μm in diameter, so that the injection of particles greater than this size will result in their being trapped in the first capillary bed that they meet. For example, when particles greater than 10 μm are injected intravenously, the particles will be trapped in the capillary beds of the lungs, thereby forming the basis for lung scanning agents (for example, albumin microspheres and macroaggregates).[5] Other organs can be visualized in a similar manner by arranging the injection to take place in a suitable vessel. The same procedure can be used to study blood flow and perfusion defects for organs such as kidneys, liver, and brain.[2]

The capillary blockage technique can also be used for thrombus localization in radioisotope venography.[6]

Table 1
RADIONUCLIDES USED IN DIAGNOSTIC IMAGING WITH COLLOIDAL PARTICLES

Radionuclide	Half-life (T50%)	Principal photons (keV)
^{51}Cr	27.8 days	320
^{57}Co	270 days	122
^{75}Se	118.5 days	136
		265
^{99m}Tc	6.0 hr	140
^{111}In	2.8 days	171
		245
^{113m}In	1.7 hr	393
^{123}I	13 hr	159
^{131}I	8.05 days	360
^{198}Au	2.7 days	412

Table 2
THE BIOLOGICAL BASIS OF LOCALIZATION OF COLLOIDAL PARTICLES AS SCANNING AND DIAGNOSTIC CONTRAST AGENTS

- Compartmental localization
  - Lymphatic system
  - Gastrointestinal transit
  - Blood flow
- Capillary Blockade
  - Lung (i.v.)
  - Kidney, liver, brain (i.a.)
  - Thrombus and thrombophlebitis
- Phagocytosis
  - Liver
  - Spleen
  - Bone marrow
  - Tumors, abscesses, lymph nodes

3. Phagocytosis

Endocytosis is the term used to describe the uptake into cells of extracellular material and it can be subdivided into two separate processes, phagocytosis and pinocytosis. Phagocytosis is the cellular engulfment of particulate materials (colloids), whereas pinocytosis is the process by which macromolecules and dissolved materials can enter the cell when the cell membrane is impermeable to such materials.

The body has many types of phagocytic cells in the normal and pathological state (for example certain tumor cells can engulf colloidal particles). In particular, the reticuloendothelial system (RES) is made up of different types of phagocytic cells, the most active of which are the macrophages. These macrophages originate in the bone marrow and can be fixed or mobile. A major function of fixed macrophages is the trapping and degradation of colloidal material and these cell types are found in the liver, spleen, bone marrow, and lymph nodes. The spleen and liver contain 85 to 90% of the reticuloendothelial cells of the human body; the Kupffer cells of the liver being a major site for particle uptake.[7] Thus, it is to be expected that a colloidal system administered by injection will be largely removed by the cells of the reticuloendothelial system and the phenomenon can be exploited for the imaging of the liver, spleen, bone marrow, etc. However, it should be remembered that endocytosis can occur in other types of cell including certain epithelial cells, liver parenchymal cells, and muscle cells. The contribution of these cells to overall particle uptake has been thought to be very small, but recent work by Praaning-van Dalen et al.[8] suggests that the endothelial and parenchymal cells of the liver may have a significant contribution to particle uptake for certain diagnostic colloids.

The uptake of particles by the cells of the RES often requires a process of modification by certain blood components that will be adsorbed to the particle surface and thereby promote the capture of the particle by phagocytic cells.[7] This so-called "opsonic" process will be discussed further.

C. Labeling with Radionuclides

The methods of labeling colloidal particles with various radionuclides are considered in detail in reviews and texts dealing with radiopharmacy and diagnostic imaging,[2,9-11] and certain examples will be discussed in detail for specific colloidal systems.

In most analytical and scanning procedures it is only the activity of the label that is followed, the assumption being that label and particle are to be found together. Detailed in vitro and in vivo tests need to be undertaken to ensure that this is indeed the case otherwise invalid results can be obtained.[12,13] Reports of microspheres being found in the brain, cerebrospinal fluid, and colloids crossing the gastrointestinal tract in significant quantities need to be questioned in terms of label integrity. Familiar scanning agents such as ^{99m}Tc sulfur colloid may not be as stable as originally supposed since double labeling studies (^{99m}Tc and ^{35}S) indicate that technetium can dissociate from the sulfur.[14,15]

II. SPECIFIC ORGANS

A. Lung

Colloidal particles can be used to study the lung either by intravenous administration (lung scanning) or by nebulization (lung deposition and clearance).

Particles in the size range 10 to 20 μm in diameter will be filtered out by the lung capillaries, with high extraction and primarily to the alveolar segments.[5,16] The human lung is believed to have about 2.8×10^8 alveoli and 2.8×10^{11} capillary segments. A wide variety of colloidal systems has been examined as lung scanning agents including ^{198}Au adsorbed onto carbon, ^{111}Ag and ^{203}Hg adsorbed onto ceramic materials, ^{99m}Tc-terric hydroxide aggregates as well as more recently macroaggregates and microspheres made from human serum albumin.[2,5] Today, metabolizable albumin microspheres labeled with ^{99m}Tc, ^{111}In, or ^{113m}In are the systems of choice.[13] A picture of a lung scan is given in Figure 1. Davis[5] has indicated that the size range 13.5 ± 1.5 μm diameter would be ideal, however this was too narrow for commercial production and he has proposed a range 10 to 20 μm (15 ± 5 μm). Such systems have a large margin of safety[5] provided the administered dose (particle number) is very small in comparison to the number of capillary segments — the toxicity being related to the size; large particles (90 μm in diameter) being more toxic than smaller ones.[17]

The rate of breakdown of albumin (either in the form of macroaggregates or microspheres) is dependent on the so-called "hardness" of the system which is related to the method of preparation and treatment (heating temperature, cross-linking agents etc). Total clearance from the lung field is generally seen within 24 hr but the rate of breakup of the microsphere and the loss of the label do not necessarily occur at the same rate.[5]

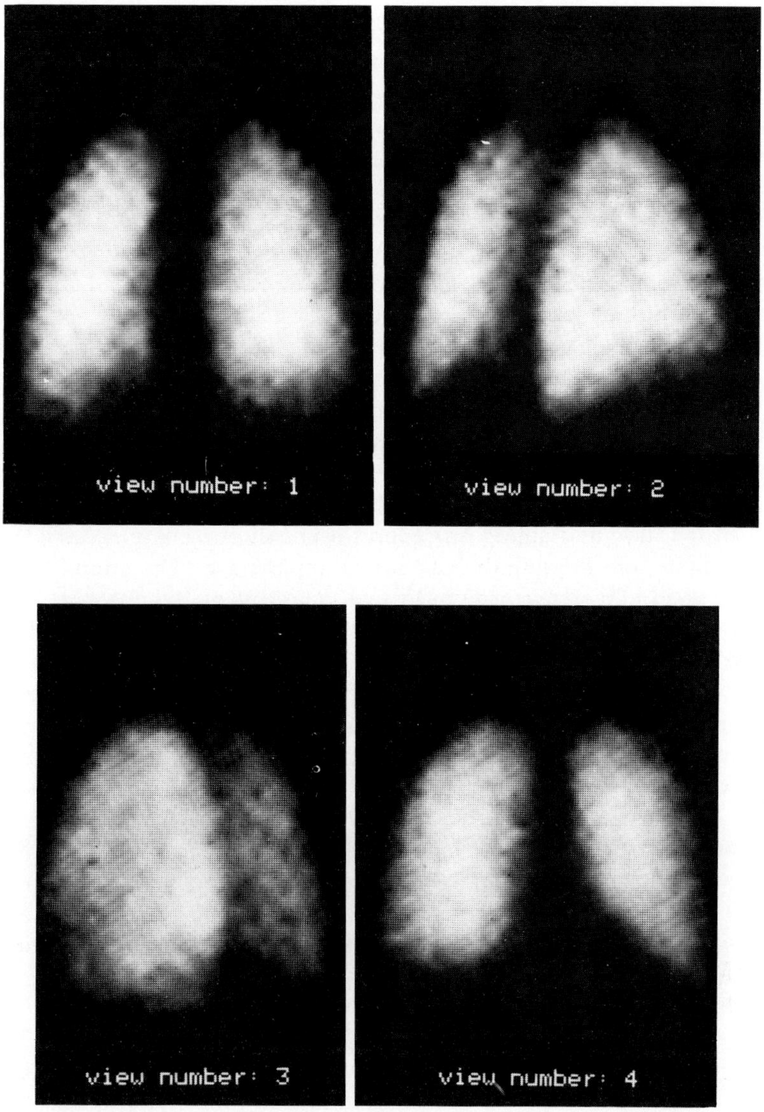

FIGURE 1. A scintiscan of the lung following i.v. administration of ^{99m}Tc macroaggregated albumin to a human subject.

Illum et al.[18,19] have used polystyrene particles surface labeled with ^{131}I and ion exchange (DEAE) cellulose microspheres and fibers labeled with ^{131}I-rose bengal to investigate effects of particle size and particle shape on lung deposition. Polystyrene microspheres of 15.7 μm mean diameter and DEAE-cellulose microspheres 40 to 160 μm diameter at a dose of about 10^8 particles were rapidly taken up by the lungs and the particles were lodged in the lung capillaries. Cellulose fibers (nominal 5 μm in length) were fatal causing blockage of the main vessels to the lungs.

Microspheres can also be administered to the lungs by the inhalation of aerosol systems.[20,21] Small albumin microspheres (1 μm diameter) labeled with ^{99m}Tc for inhalation scanning and lung clearance studies have been described by Yeates et al.[22] and by Fazio and Larfortuna.[23] The latter investigated the effects of inhaled drug (Salbutamol) on the mucociliary clearance in patients with chronic bronchitis. Monodisperse Fe_2O_3 particles labeled with ^{99m}Tc have been used for similar purposes.[24] Nonbiodegradable polystyrene and Teflon particles generated as monodisperse systems by spinning disc generator and labeled with ^{99m}Tc are now used widely in studies on lung clearance by mucociliary and cough mechanisms[25-28] and the effects of drugs and disease states. Labeled liposomes administered to the lungs are deposited in the bronchioles and alveoli and are cleared slowly.[29]

B. Liver (Spleen and Bone Marrow)

The liver and spleen contain 80 to 95% of the reticuloendothelial cells of the body (liver 80%, spleen 10%)[7,30,31] and for foreign particles it is to be expected that 80 to 90% of administered particles will be removed per blood passage through the liver[31] provided that the system is not thereby saturated or blocked. If hepatic uptake is impaired or blocked then other cells of the RES have an opportunity to take up particles (lung, spleen, bone marrow).[7]

The process of phagocytosis of colloidal particles by the cells of the reticuloendothelial system in the liver and spleen has been the subject of many reviews.[32] Those by Saba[7] and Walters and Papadimitriou[33] are recommended. The phases of uptake include opsonization by plasma components, attachment of the opsonized particles to phagocytic cells, phagocytosis, and particle degradation. Not all these processes are essential for uptake. Some particles such as polystyrene are apparently taken up by Kupffer cells without opsonization and of course cannot be degraded.[33] The processes of opsonization and subsequent uptake are highly specific and involve various surface phenomena such as surface charge and surface hydrophobicity. It is not known whether the opsonin recognizes the particle or is attracted to the particle as a random event. Saba[7] has suggested that the conformation of the opsonic material at the particle surface renders the surface foreign and thereby recognized. More than one opsonic system is in operation. Some opsonins will lead to binding to phagocytic cells without internalization.[33] Apparently the phagocytic receptors can function independently of each other and as a result blocking the RES with one colloid does not necessarily prevent the uptake of another colloid.[7]

The accumulation of particles in the liver will be determined by:[7,34]

- Liver sinusoidal blood flow
- In vivo opsonization
- Size and surface of test particle
- Receptor density in liver
- Number and phagocytic activity of Kupffer cells

The reticuloendothelial system can be altered by pathological conditions as well as by administered drugs and other substances (shock, tumor growth, bacterial and viral infections, atherosclerosis, radiation, immunological suppression). Glucan, Zymosan, estrogen, endotoxin, and Bacillus Calmette Guerin are all stimulants of the RES, while cortisone, methyl and ethyl palmitate and antilymphocyte serum (ALS) depress the RES.[7]

The colloidal properties that affect uptake by the RES are size, dose, charge, and nature of the surface.[7,18,32] Large particles are removed more quickly than small ones and the smaller particles tend to concentrate in bone marrow and spleen[32,35,36] (Figure 2). The charge on the particle can affect the processes of opsonization and subsequent adhesion between particle and phagocytic cell.[37,38] Similarly, surface properties are largely responsible for opsonization and uptake. Various authors have proposed factors such as surface roughness and surface hydrophobicity as important variables[35,39]

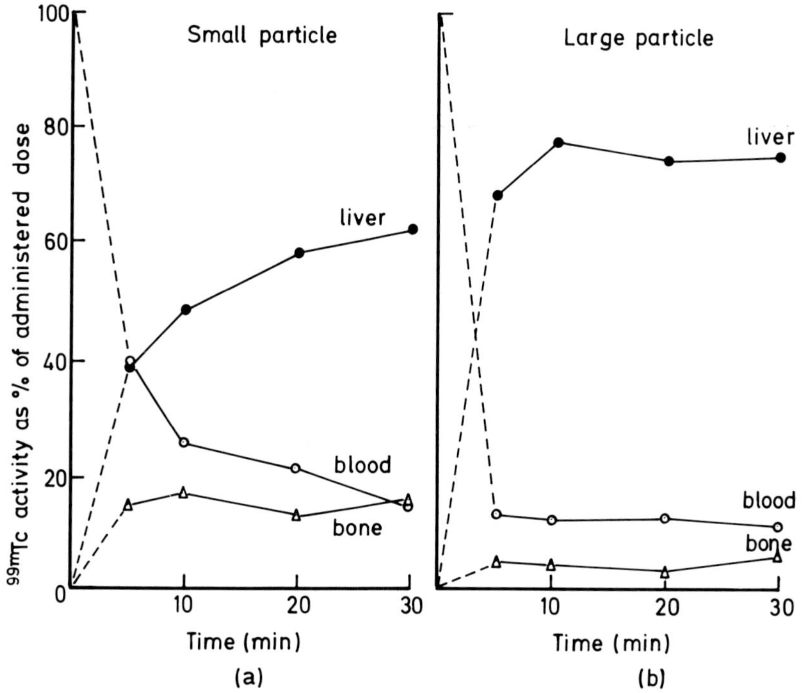

FIGURE 2. The effect of particle size on the distribution of antimony sulfide colloid following i.v. injection.

but these have been difficult to determine experimentally (see below). The coating of colloidal particles with nonionic surface active agents and polymers can have a profound effect on liver uptake and the redirection of particles to other sites[36,38] (Figure 3).

The ideal properties of a radiodiagnostic material for studying the reticuloendothelial system administered in the blood circulation have been outlined by Benarcerraf et al.[31] and Colombetti.[40]

- Be taken up only by the macrophages in contact with blood
- Not cross into extravascular space during the test period
- Be homogeneous in size
- Be stable in blood, in the concentrations used, exhibit no loss of radiolabel or alteration of particle size
- Be nontoxic for macrophages and patient
- Be stable in vitro as well as in vivo
- Be accurately measurable
- Give low radiation dose to patient

Systems that have been used in radiodiagnostic imaging include:[1,2]

- ^{99m}Tc-sulfur colloid
- ^{99m}Tc-antimony sulfide colloid
- ^{113m}In-gelatin colloid
- ^{198}Au-colloid
- ^{99m}Tc-stannous phytate
- ^{131}I-albumin microaggregates
- ^{99m}Tc-albumin microspheres

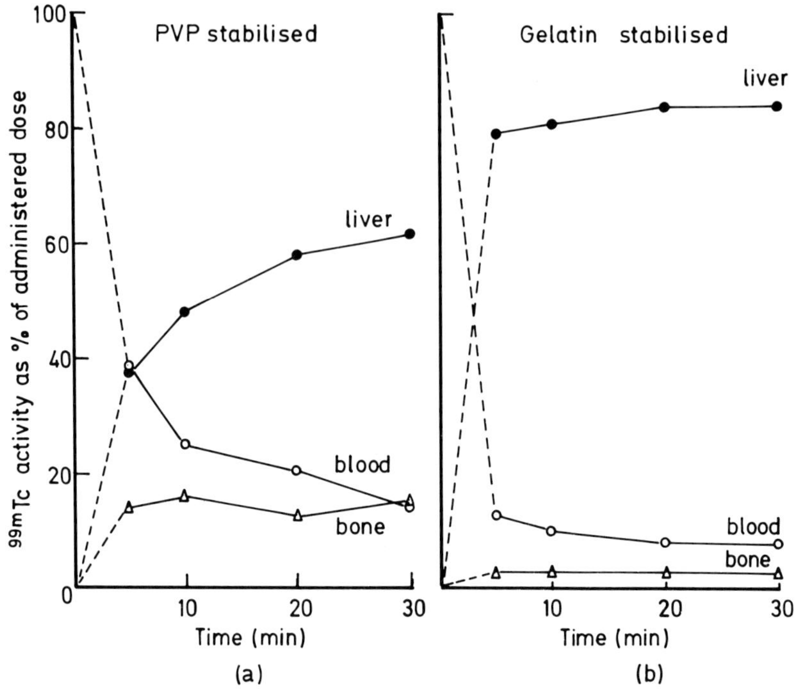

FIGURE 3. The effect of stabilizing agents on the distribution of antimony sulfide colloid following i.v. injection.

Much of the early imaging work was performed with gold colloid ^{198}Au. This has now been completely superceded by ^{99m}Tc-labeled colloids, particularly sulfur colloid, antimony sulfide colloid, and millimicrospheres of denatured human serum albumin. With all these materials a high proportion of the injected dose reaches the liver within a few minutes of administration. The rates and extents of uptake are not the same, thereby indicating the physicochemical differences which exist between the various systems. Clinically, sulfur colloid labeled with the radionuclide ^{99m}Tc remains the most commonly used material for liver and spleen imaging. The reason for its preference over antimony sulfide colloid is largely one of ease of preparation.[1] It tends to be preferred to millimicrospheres because of some instability of labeling shown by the latter. Technetium is released in vivo within a relatively short time following administration, contributing to a high proportion of bloodborne activity and a degradation of image quality. The factors affecting the distribution of ^{99m}Tc sulfur colloid stabilized by gelatin have been discussed by Atkins et al.[41] who concluded that particle size and numbers were both of importance in determining the distribution of the colloid in the reticuloendothelial system. Blocking agents such as gelatin itself, polystyrene latex, and carbon affected the distribution. Liver and spleen scans obtained with ^{99m}Tc labeled sulfur colloid are shown in Figure 4. Certain other preparations have found their way into clinical use, e.g., tin colloid and the phytates. In these systems the colloidal particles are formed in vivo and are poorly defined, and the preparations will not therefore be further discussed. Questions have also been raised about the specificity of albumin microspheres for Kupffer cells[8] (Table 3). However, the albumin millimicrosphere system is still popular since it is readily available in kit form and can be labeled in a one-step process.[42]

^{99m}Tc-labeled liposomes of small size 10 to 500 nm have also been proposed as liver scanning agents but once again questions of label integrity have been raised.[43,44] Lipid

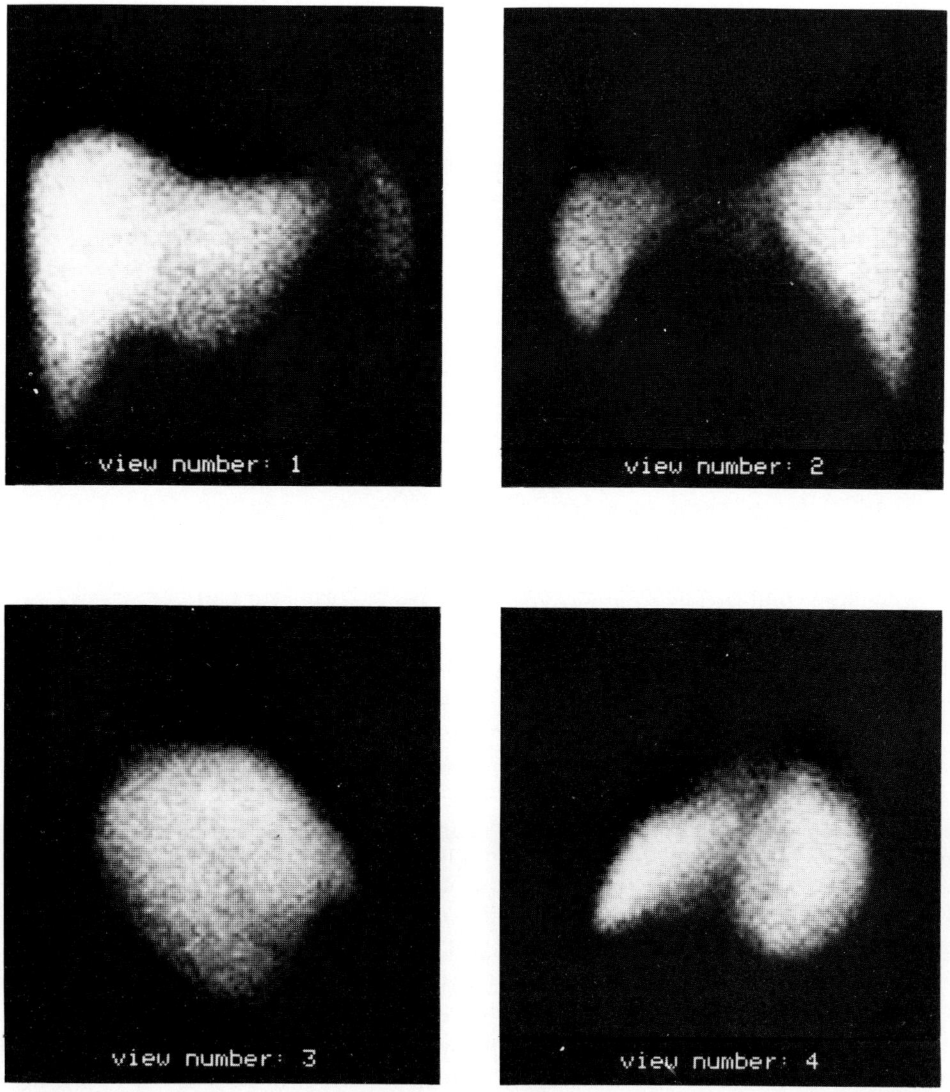

FIGURE 4. A scintiscan of the liver and spleen following i.v. administration of ^{99m}Tc sulfur colloid to a human subject.

soluble ^{99m}Tc-labeled compound materials, or chelating agents, are now used to provide greater stability.[29] As with microspheres, size, composition, and charge are all important variables that determine site and extent of deposition. It has even been claimed that positively charged liposomes can reach the brain.[29] Empty liposomes have been used to block the RES of the liver and thereby divert labeled liposomes to tumor sites.[45,46] Gelatin particles (nanoparticles) have been labeled with ^{99m}Tc by Oppenheim[47] using the method of Eckelman et al.[48] The particles were taken up rapidly by the liver but there was rapid loss of the radioactivity, either by degradation of the particle or loss of the label.

The kinetics of uptake (and clearance) of colloidal particles by the liver have been considered[30,34,49] using nonlinear and compartmental models. Boisvieux et al.[49] used

Table 3
LOCALIZATION OF TEST SUBSTANCES IN DIFFERENT TYPES OF LIVER CELLS[8]

Colloid	Cell type: Parenchymal	Kupffer	Endothelial
Colloidal albumin (ng/10^6 cells)	190 ± 38	212 ± 97	624 ± 217
Antimony sulfide	ND[a]	493 ± 67	259 ± 60
Colloid (pg/10^6 cells)	910 ± 320[b]	2800 ± 930	1380 ± 30

Note: ND, not determined.

[a] Low dose.
[b] High dose.

^{198}Au-colloid stabilized by gelatin to determine the number of Kupffer cells and rate of phagocytosis in rats. Reske et al.[34] proposed a three-compartment model to evaluate monoexponential accumulation of ^{99m}Tc-HSA microspheres in Kupffer cells of patients, followed by a two-stage elimination process. ^{99m}Tc sulfur colloid was used by Ryder et al.[30] to assess the RE function in normal and tumor bearing animals using a two-compartment model. The kinetics of uptake of particles by the RES in tumor subjects is different to that for normal individuals.[30,50]

Emulsion systems made from iodinated vegetable oils can be used as contrast media in radiology, and ever increasingly in computed tomography (TC) as contrast materials.[51-53] These authors have described their experiments with an emulsion formulated from an iodinated ester of poppy seed oil. Interestingly, they reported that in animal experiments the emulsion was largely in the Kupffer cells of the liver 1 hr postadministration. However, at 6 hr, the emulsion was equally distributed between Kupffer cells and hepatocytes and after 24 hr the emulsion was almost totally within the hepatocytes. Such results have implications for fat emulsions used in parenteral nutrition as well as for drug targeting.

In a similar type of study Laval-Jeantet et al.[54] have studied particle size effects following the intravenous administration of iodinated fat emulsions once again intended as CT contrast media. Three different sizes were examined, 7.0, 1.3, and 0.7 μm average diameter. The coarse emulsion was taken up by the liver and lung while the fine and ultrafine emulsions were found in the liver and the spleen. The ultrafine emulsion was metabolized very quickly and free iodine was released rapidly (Table 4). The authors suggested that the fine emulsion (1.3 μm) was fixed in the liver sinusoids by a process of embolization analogous to the uptake in the capillary beds of the lung, whereas the ultrafine emulsion was removed by the Kupffer cells and metabolized. They offered no direct experimental evidence for their proposals and their views are contrary to accepted theories of RE uptake. Davis and others[55-57] have used iodinated fat emulsion to investigate the uptake of lipid emulsions by the liver and spleen and the role of the nature of the emulsifier. Nonionic emulsifiers led to a different pattern of behavior than did phospholipid (lecithin) systems. Growing interest in the use of emulsions and liposomes will no doubt occur with developments in the field of NMR imaging. Fluorocarbon emulsions used presently as red blood cell substitutes[58] could be used as contrast agents. Similarly, liposomes that can penetrate cell membranes of

Table 4
EMULSION DEPOSITION AND THE EFFECT OF PARTICLE SIZE (IODINATED EMULSION)[54]

Size (μm)	Deposition (%) at 3 hr (mean ± SD, n = 15)		
	Lung	Liver	Spleen
7 (± 4)	45 ± 12	48 ± 8	4 ± 1.7
1.3 (± 0.02)	6 ± 3	63 ± 7	2 ± 1.2
0.7 (± 0.2)	1.0 ± 0.4	12.5 ± 4	27 ± 8

specific cells to deliver substances to modify the cellular milieu could produce biochemical changes detectable by NMR.[29]

Heavy metal contrast media for use in CT of the liver has been examined by Havron et al.[59] These included silver iodide colloid and oxides of the elements in the range 58 to 66 since these absorb most X-rays from a 120 kVp energetic beam (cerium, dysprosium, gadolinium). The systems were better than iodinated systems as contrast agents but had a long dwell time in the liver (and spleen). The colloidal oxides needed to be stabilized by protective materials such as polyvinylpyrollidone, albumin, and gelatin in order to give stable systems of an appropriate size.

C. Lymphatic System

The injection of colloidal particles into the interstitial spaces (via intradermal, subcutaneous, intramuscular injection) causes them to reach the lymph system via the lymphatics that drain the site of injection. The distribution of the colloid in the lymphatics and particularly in the lymph nodes will reflect the physiological and pathological state of the lymphatic system.[4,60,61] The uptake of particles in the nodes is a result of a combination of filtering (entrapment) and phagocytosis[62] by cells of the reticuloendothelial system. The uptake is not absolute, instead continuous slow transport of activity through the lymphatic system takes place as does progressive accumulation of activity at all nodes.[4] The particle size of the colloid is of critical importance in determining transport and uptake by the RE cells.[63] The optimum size is in the range 10 to 150 nm and shape has also been implicated.[6] Strand et al.[63] have found recently that a lower figure of 5 nm is more applicable. In their studies they injected ^{99m}Tc-sulfur colloid (600 nm), ^{99m}Tc-antimony sulfide colloid (5 to 10 nm) and ^{198}Au colloid (5 nm). The last system migrated well to the lymphatics. A comparison of ^{99m}Tc-antimony sulfide colloid and the *in situ* colloid agent ^{99m}Tc-stannous phytate has been made by Ege and Warbick.[64]

The use of ^{99m}Tc-labeled antimony sulfide colloid and sulfur colloid to study lymph nodes in various tissue sites has been reported by Ege[65,66] and Aspegren et al.[67] who found good correlation between scintigraphic data and microscopic examination of nodes removed at operation. The effect of particle size was once again stressed. Much of the published work describing the imaging of lymph nodes for clinical purposes has been performed using antimony sulfide colloid.[65] Other systems (e.g., rhenium sulfide) have been less widely utilized. As with liver imaging, much of the early work was performed with gold colloid (^{198}Au). This has now been discontinued largely because of considerations of radiation dose to the patient. A patient study showing lymph nodes of the internal mammary chain is shown in Figure 5.

Liposome systems have also been used for lymph node visualization.[29,44,68] Wu et al.[69] reported different distribution characteristics for liposomes that were dependent

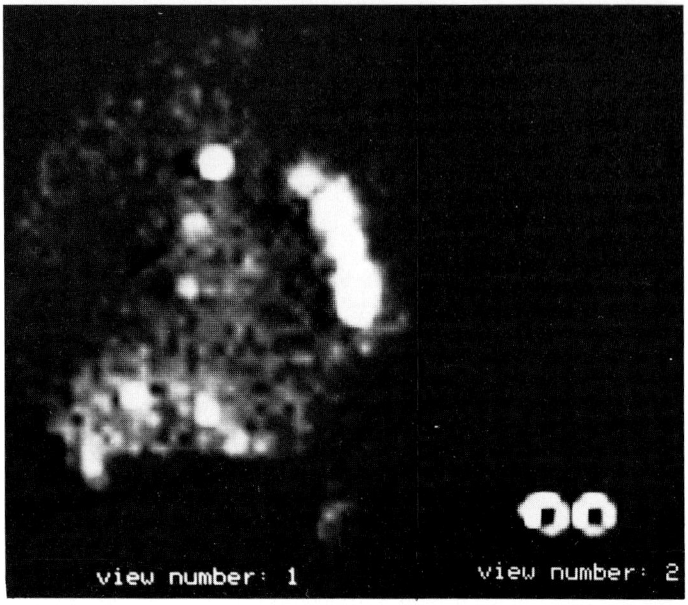

FIGURE 5. A scintiscan showing lymph nodes following s.c. injection of ^{99m}Tc antimony sulfide colloid to human subjects.

on surface characteristics. Liposomes containing glycolipids injected into the mouse foot pad, and by sc administration had a different distribution to those without the glycolipid.

D. Blood Flow Determination (Tracer Microspheres)

So-called "tracer microspheres" available originally from the 3M Corporation and now from New England Nuclear are polymeric particles of a size 15 ± 5 μm diameter that have been used widely to follow blood flow and other physiological processes in animals (and where applicable in man) (airways clearance, genital tract migration, ovum transport). The particles are labeled with radionuclides such as ^{85}Sc, ^{141}Co, ^{103}Ru, ^{46}Sc, ^{57}Co, ^{51}Cr etc. Their use in blood flow measurements has been reviewed by Bartium et al.[70] and Chan et al.[71] In addition, in 1980, the 3M Company published a comprehensive bibliography of over 800 papers dealing mainly with regional blood flow measurements, cardiac output, and shunt determinations (arteriovenous anastomosis).[72] In blood flow determinations the particles are injected into the artery in question and the microspheres are presumed to distribute according to the precapillary distribution of blood flow through the microcirculation. The particles are trapped in vessels of similar diameter to themselves and thus do not enter the venous circulation.[73]

Particle size is an important factor in trapping efficiently. For example if the particles are greater than 15 μm diameter then no transport through the lungs is observed, while if they are less than 10 μm in diameter 20% can pass through the lung capillary beds.[74] Segadal and Svanes[75] evaluated the microsphere method for measuring cardiac output and found it to be suitable for small and medium-sized animals. Tepperman and Jacobson[73] have concluded recently that the microsphere method is comparable to other techniques for the determination of blood flow. Determination of coronary blood flow may be performed using this technique, by direct injection into the coronary artery. The method has been termed particulate myocardial perfusion scintigraphy (PMPS)[76] and typically requires the injection of 20,000 particles. Again the important properties are those of particle size and the ease with which the particles may be degraded.

The effect on hemodynamics of single and repeated administration of 15 ± 5 μm microspheres has been investigated.[74,77,78] No effects on blood pressure, cardiac output, or regional blood flow were observed for an injection of three separate infusions of 10^5 particles[77] and Heymann et al. have recommended that for most species 3 to 5 × 10^5 particles can be used with no adverse effects. Smaller particles (3.4 μm mean diameter) which can pass through the lungs cause hypotension.[78] Tracer microspheres can also be made from ceramic materials, dextran, and albumin.[74] Recently there has been interest in using degradable albumin microspheres for blood flow determination using ^{99m}Tc and ^{113m}In labels for scintigraphic evaluation[79] and ^{11}C for positron tomography.[80] Kanke et al.[16] used scintigraphic methods to follow the clearance of ^{141}Ce labeled tracer microspheres after i.v. administration. The commercially available microspheres were made from divinyl benzene and were from 3 to 12 μm in mean diameter. The deposition at organ sites (lung, liver) was dependent on particle size and hence the clearance mechanism (capillary entrapment or reticuloendothelial cell uptake). Madden et al.[81] have presented some unusual results describing the recovery of microspheres from washings from the oropharynx, trachea, bronchi, stomach, and bladder as soon as 15 sec after injection of particles into the right atrium of the heart or the aorta. Heymann et al.[74] could not duplicate these findings and suggested that a radiolabel was being lost rapidly from the microspheres in Madden's study. However, this explanation is not valid since Madden et al. used a microscopic method to detect their particles. The mechanism for such rapid transit to extravascular sites (if not an experimental artefact) is not clear.

E. Gastrointestinal Tract

Microspheres based on ion exchange resins of a size in the region of 1 mm have been used very successfully to study gastric emptying and intestinal transit in animals and man as well as for imaging segments of the gastrointestinal tract.[82-84]

Various ion exchange materials have been studied and the release of label into gastric and intestinal fluids has been measured. Commercial resins bearing strong anion exchange functions can be easily labeled with ^{99m}Tc. Wirth et al.[85] described ^{99m}Tc-labeled systems based on styrene-divinylbenzene copolymers with paired iminodiacetate chelating groups that had a strong affinity for transition metals. A wide range of ion-exchange systems has been examined by Theodorakis et al.[86], using both commercial and their own resin systems. A variety of acceptable materials were reported (Table 5) that released little of the label under physiological conditions and did not exchange with food particles or the stomach wall.[87] Hinder and Kelly[88] have used much smaller microsphere systems (tracer microspheres) to follow the gastric emptying of solids and liquids in the dog. These systems have also been employed as fecal markers. ^{99m}Tc-sulfur colloid has been used as a marker to follow gastrointestinal transit in rat of different oily formulations.[89,90]

III. MEASUREMENT OF PHYSICOCHEMICAL PROPERTIES OF PARTICLES

The foregoing discussion has shown that the important physicochemical properties of colloidal particles have a direct effect upon the extent of distribution and deposition in the body. The following are identified:

- Particle size and size distribution

Table 5
THE BINDING AND RELEASE OF $^{99m}TcO_4^-$ ANIONS BY POLYSTYRENE MICROSPHERES [86]

Polymer	Paricle size (mesh)	% Binding (3 mCi, 60 min) pH 1.5	5.0	8.0	Release: In vitro % release gastric fluid	In vitro % release intestinal fluid
P-Ø (polystyrene)	40—100	< 10	< 10	< 10	—	—
P-Ø-TETA (P-TETA)	40—100	95.2	98.9	98.1	3.0	7.2
CH_2CH_2OH P-Ø-CH_2N^+-$(CH_3)_3$ (Dowex 2-X8)	50—100	98.5	99.7	99.9	1.1	0.8
P-Ø-CH_2N^+ $(CH_3)_3$ (AGI-X2)	200—400	96.3	98.6	99.7	4.6	2.6
P-Ø-CH_2-CH-CH_2-$N(CH_3)_2$ CH_2-CH_2-CH_2 (Bio-Rex 9)	200—400	95.3	99.3	99.5	4.4	3.8
P-Ø-$CH_2NH(CH_2COOH)_2$ (Chelex 100)	100—200	92.2	88.5	90.1	—	—
P-Ø-SO_3H (Dowex 50W-X8)+ $SnCl_2$ $2H_2O$	100—200	93.1	68.0	71.1	—	—

Note: Values are means for three replicates. Ø — aromatic ring of the polystyrene resin.

- Particle shape
- Surface charge
- Surface nature (to include factors such as roughness, hydrophobicity, and affinity for blood and tissue components)
- The nature of any of the added surface coating agents

Some of these factors have been examined in great detail (e.g., particle size) whereas others are mentioned as being important but little studied in systemic investigations. A major problem has been the difficulty in measuring the colloidal properties of interest.

A. Particle Size

The range of methods available for particle size analysis has been reviewed.[90] Large particles intended for studies on gastrointestinal transit can be measured by microscopy or sieve analysis. Microscopy can also be used to measure particles in the size range 2 μm and above. Particles of a size 800 nm to 20 μm are best evaluated using an electronic counter such as a Coulter Counter, fitted with a sampling tube of appropriate size. The Coulter Counter is capable of measuring sizes of less than 800 nm but in practice the method is subject to electrical interference. Electron microscopy and chromatographic methods[63] have been used for submicron systems. While these methods are suitable they can be tedious to perform and are not entirely suitable for routine quality control purposes. Particle size analysis results for sulfur colloid obtained using different methods are given in Table 6. Laser light scattering methods using photon correlation techniques are rapid and provided that the particle size is unimodal and free from large interfering material can be used to provide both mean diameter and polydispersity extremely rapidly.[92] The method has been used with success for a variety of colloidal materials used as radiodiagnostic agents.[93]

Table 6
PARTICLE SIZE AND NUMBER IN SULFUR COLLOID DETERMINED BY DIFFERENT METHODS

Method	Particle size (μm)	Particle number ($\times 10^{-8}$ mℓ^{-1})
Autoradiography	—	5.5 (SD ± 1.5)
Nucleopore filtration of ^{35}S colloid	0.4—0.6	9.1—2.7[a]
Electron microscopy	0.5	4.7[a]

[a] Calculated from particle size data based on the following assumptions: particles are spherical; sulfur has a density of 2 g/cc; the yield of colloidal sulfur from the outer S atom of thiosulfate is 50%.

B. Particle Shape

Although particle shape has been listed as a variable affecting uptake by phagocytes[7] the literature apparently contains no detailed studies. Illum et al.[18] investigated the effect of particle size and shape on lung deposition for cellulose particles of 5 μm and larger. With smaller particles it is difficult to change shape without modifying other physiocochemical variables.

C. Surface Charge

Colloidal particles will have a charged surface, either through a process of ionization of surface groups or groupings from the material itself or from coating agents, or the adsorption of ions from solution. Even particles coated with so-called nonionic polymers (surface active agents) will carry a small negative charge. Particles can be designed to have charges of different size and magnitude[94,95] either by incorporation of ionized or ionizable species during preparation of the colloid or by surface adsorption after preparation. There is some evidence to suggest[96] that positively charged particles have a higher affinity for certain tissues (lung, heart) than similar negatively charged particles. However, it is difficult to change charge without also modifying the nature of the surface itself. It must also be remembered that once the particle is in contact with body fluids it will rapidly acquire a negative charge through the adsorption of one or more of a variety of proteins, immunoglobulins, or glycoproteins.[97] The charge that the particle carries in the body (and the nature of the coating) will be relevant to its fate. Thus, the initial charge state will affect processes that occur after administration. Surface charge measurements can be carried out in saline, at different pH values and in the presence of plasma and serum in order to gain insight into the nature of ionizable groupings before and after administration.

The technique of cell microelectrophoresis can be employed for particles of 500 nm and above.[98] Here the particles are placed in an electrical field and the mobility measured by microscopic observation. With smaller particles this method is no longer applicable and instead more sophisticated techniques such as laser doppler anemometry are required. The principle is also one of determining mobility in an electrical field, the method of measuring that mobility relies on a laser technique and correlation methods rather than optical observation.[99] Our preliminary evaluation of this method shows it to be rapid and accurate and the derived data are in good agreement with measurements made with conventional microelectrophoresis.

D. Surface Affinity

The nature of the colloid surface will determine whether it interacts with blood and tissue components. Van Oss et al.[39] have proposed that surface hydrophobicity is the all important factor; namely that hydrophobic particles require no opsonization and will be taken up avidly by cells of the reticuloendothelial system, while more hydrophilic particles will need to be made more hydrophobic by opsonization in order for them to be recognized and removed. Van Oss et al.[39] have used contact angle measurements as a measure of surface free energy or surface hydrophobicity. This technique is difficult to apply to small colloidal particles. However, contact angle measurements with plane surfaces made from the same material as the colloid can be made.[100] The surface hydrophobicity of particles can also be related to surface roughness and used to explain the difference between pathogenic and nonpathogenic strains of bacteria.[101,102] Other methods of measuring surface hydrophobicity include two-phase partition of particles using glycol dextran mixtures,[103] hydrophobic affinity chromatography,[104] the adsorption of radiolabeled (^{14}C) fatty acids and alkanes to particles[105] and fluorescent probes.[106] Each method has its advantages and disadvantages and in our own hands no one technique stands out so far as being the best. For example, methods based on chromatographic techniques are suitable for studying differences between particles of the same average size and distribution but are difficult to apply if the particles have different surfaces and size characteristics.

E. Surface Coatings

The adsorption of selected proteins and blood components can be studied in vitro using labeled proteins (^{125}I) and measurement of adsorption isotherms in the usual way. Similar adsorption isotherm experiments can be conducted using coating agents intended to influence uptake of particles by the cells of the RES.[107]

The manner in which the final colloidal particle will interact with phagocytic cells can be evaluated in vitro using models such as mouse peritoneal macrophages[108] or the more recently developed isolated in vitro Kupffer cell method.[109] The rate and extent of uptake can be followed by particle counting, turbidometric or radiometric methods. The chemiluminescence that occurs during the phagocytic process may provide another possible approach.[33]

IV. SELECTED RADIODIAGNOSTIC COLLOID SYSTEMS

In order to illustrate the preparation of some different types of colloidal particles for radiodiagnostic use, details are given for four systems; sulfur colloid, antimony sulfide colloid, albumin millimicrospheres, and polystyrene microspheres.

A. Sulfur Colloid

Particles of colloidal sulfur are commonly prepared by the reaction between sodium thiosulfate and hydrochloric or phosphoric acid in the presence of a protective colloid such as gelatin. Several authors have examined the properties of sulfur colloids. Davis and his co-workers[110] introduced the use of polycarbonate membrane filters as a rapid technique for particle sizing. Frier, Griffiths, and Ramsey[15,111] examined particle size and particle number using a variety of techniques. In vivo stability of sulfur colloid has been examined using a dual isotope method in which ^{35}S particles are labeled with ^{99m}Tc. Significant changes in particle size distribution, together with the release of soluble sulfur compounds, takes place when sulfur colloid is placed in contact with solutions containing sulfide, or with serum (Figure 6).[15,111] In vivo distribution of the two radionuclides following i.v. administration of doubly-labeled colloids differs, as shown in Figure 7, indicating that the observed in vitro interactions do take place in vivo. At all times ^{99m}Tc activity remains associated with particles.

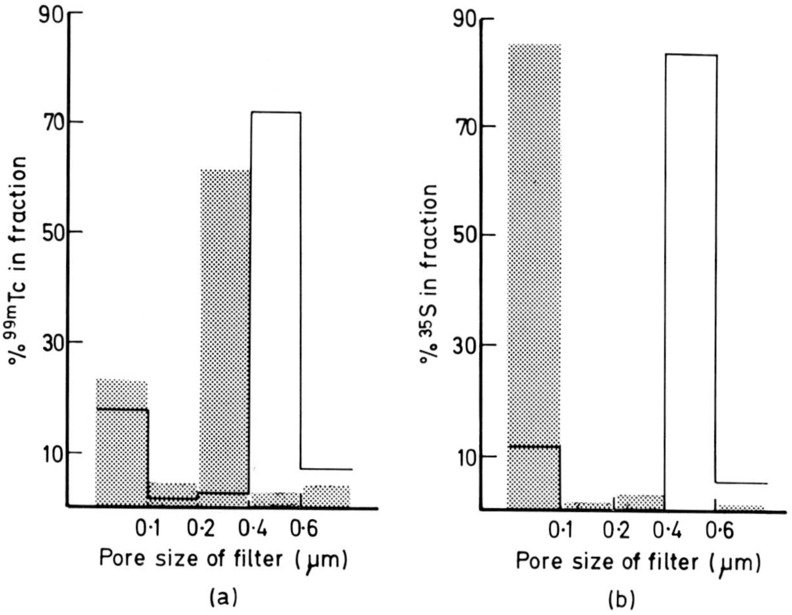

FIGURE 6. Activity/size distribution of $^{99m}Tc/^{35}S$ colloid following contact with serum (unshaded boxes, before contact; shaded boxes, after contact).

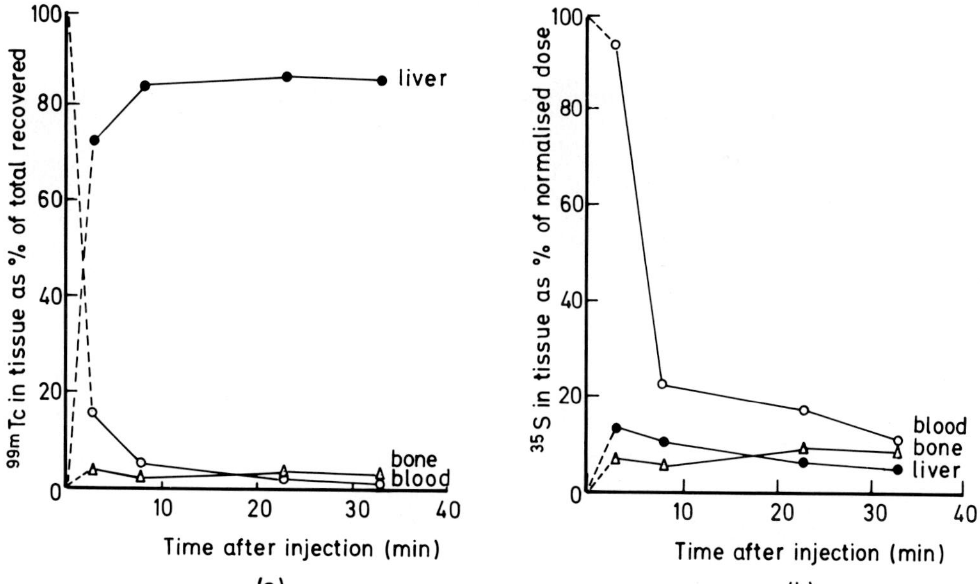

FIGURE 7. The distribution of ^{99m}Tc and ^{35}S following i.v. injection of a doubly labeled sulfur colloid.

B. Antimony Sulfide Colloid

The preparation of antimony sulfide colloid requires the interaction between antimony potassium tartrate and hydrogen sulfide at a controlled pH.[112] The resulting suspension is commonly stabilized by the presence of polyvinylpyrrolidone. Particles are considerably smaller than sulfur colloid, being generally of the order of 10 nm. Sizing requires the use of techniques such as electron microscopy[113] or photon corre-

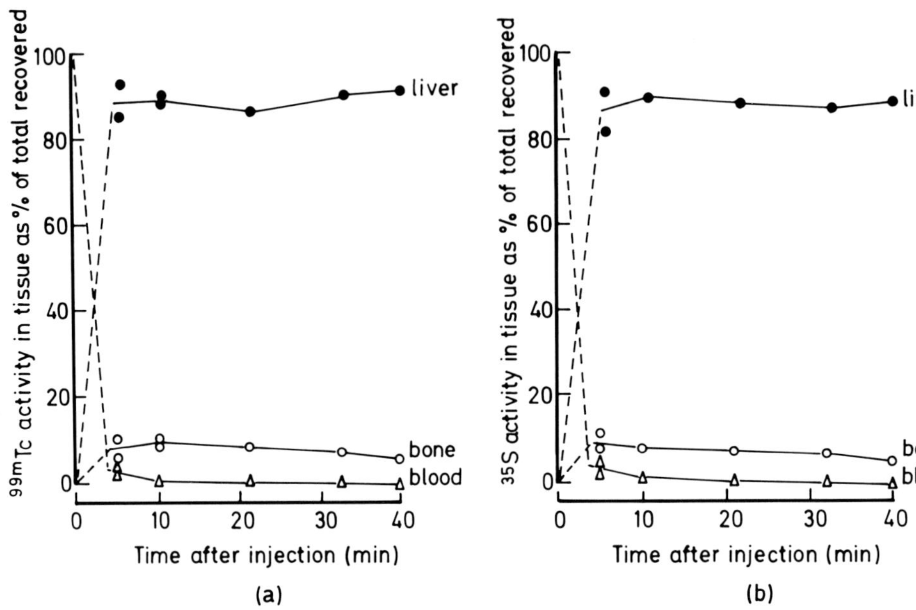

FIGURE 8. The distribution of ^{99m}Tc and ^{35}S following i.v. injection of a doubly labeled antimony sulfide colloid.

lation spectroscopy. Comparison of the in vivo distribution of antimony sulfide colloid with that of sulfur colloid shows a slower rate of clearance from the blood, a smaller proportion of activity in the liver, and an increase in the proportion of activity found in the bone marrow. Antimony sulfide is not commonly used as a liver-scanning agent but finds its most useful application in the imaging of lymph nodes.

Experiments involving the preparation of doubly labeled antimony sulfide colloids[114] have shown them to have a high degree of stability both in vitro and in vivo. Biodistribution studies show very close correlation between ^{99m}Tc and ^{35}S activities in the organs examined following administration of a colloid labeled with these two radionuclides (Figure 8).

C. Millimicrospheres

Human serum albumin, when subjected to mild heating in solution at a pH close to the isoelectric point of the protein will precipitate readily. Controlled precipitation produces a suspension of particles having known size characteristics. Commercially produced millimicrospheres are of the order of 0.5 to 1.0 μm in diameter, and the formulation may contain some surface active agent to improve the wetting characteristics and stability of the suspension. Radionuclide labeling for clinical purposes is generally accomplished by the addition of ^{99m}Tc to a suspension of performed particles, suggesting that the radionuclide undergoes surface attachment only.[48] This implies that stability of labeling will not be high, and clinical experience shows that there is indeed a release of activity in vivo from sites of uptake shortly following administration. Being proteinaceous in nature, the particles are subject to degradation in vivo by a combination of chemical dissolution and enzymatic action.[5]

D. Polystyrene Latex

Polystyrene latex particles are commercially available in sizes from around 50 nm to about 100 μm and appear as monodisperse spheres with a very narrow size distribution.

The particles can be obtained with a range of surface groupings and consequently with different surface charges. Polystyrene particles can be effectively labeled with ^{131}I when irradiated with a ^{60}Co source in the presence of ^{131}I-sodium iodide without altering the stability of the particle suspension or the particle size.[115] When irradiated in the presence of iodide, a chemical bonding of iodine to the surface occurs, presumably through the formation of C-I bonds. Since the reaction rate is proportional to the surface area small particles can be labeled very easily while it is more time consuming to label large particles.[115]

V. CONCLUSIONS

Radioactively labeled suspensions have been used for many years to image the reticuloendothelial system. The materials have been termed "colloids" although several systems are not truly colloidal in nature. Principal applications in nuclear medicine are the imaging of the liver, spleen and to a lesser extent, bone marrow, using colloids administered by the intravenous route. The imaging of lymph glands is possible, utilizing a subcutaneous method of administration. Certain colloidal systems have proved more successful than others for particular applications for reasons which include ease of preparation, stability, and biological fate, the latter being dictated largely by consideration of particle size and number, and the characteristics of the particle surface.

Larger particle systems where the individual particles may exceed 15 μm in diameter are used for lung imaging and for the demonstration of blood flow in organs such as the heart. The mechanism of uptake is purely one of entrapment due to the size of the particle exceeding that of the vessels through which the blood is flowing. Particle properties other than size have little bearing on the uptake sites, this being determined mainly by the site and route of administration. Major factors governing choice of formulation concern the stability and long term fate — particles must be sufficiently stable to allow the completion of the study, but must degrade eventually such that permanent capillary blockade, leading to embolizaton, does not result.

REFERENCES

1. McAjee, J. G. and Subramanian, G., in *Clinical Scintillation Imaging,* 2nd ed., Freeman, L. M. and Johnson, P. M., Eds., Grune & Stratton, New York, 1975, 13.
2. Solway, A. H. and Davis, M. A., Survey of radiopharmaceuticals and their current status, *J. Pharm. Sci.,* 63, 647, 1974.
3. Rhodes, B. S. and Croft, B. Y., *Basis of Radiopharmacy,* C. V. Mosby, St. Louis, 1978, 33.
4. Cox, P. H., The kinetics of macromolecule transport in lymph and colloid accumulation in lymph nodes, in *Progress in Radiopharmacology,* Vol. 2, Cox, P. H., Ed., Elsevier/North Holland, Amsterdam, 1981, 267.
5. Davis, M. A., Particulate radiopharmaceuticals for pulmonary studies, in *Radiopharmaceuticals,* Subramanian, G., Rhodes, B. A., Cooper, J. F., and Sodd, V. J., Society of Nuclear Medicine, New York, 1975, 267.
6. Cox, P. H., Particulate reagents for thrombus localisation, in *Progress in Radiopharmacology,* Vol. 2, Cox, P. H., Ed., Elsevier/North Holland, Amsterdam, 1981, 119.
7. Saba, T. M., Physiology and physiopathology of the reticuloendothelial system, *Arch. Intern. Med.,* 126, 1031, 1970.
8. Praaning-van Dalen, D., Brouwer, A., and Knook, D. L., Clearance capacity of rat liver, Kupffer, endothelial and parenchymal cells, *Gastroenterology,* 81, 1036, 1981.
9. Rocha, A. F. G. and Harbert, J. C., *Textbook of Nuclear Medicine: Clinical Applications,* Lea & Febiger, Philadelphia, 1978, 196.
10. Spencer, R. P., Ed., *Radiopharmaceuticals, Structure Activity Relationships,* Grune & Stratton, New York, 1981.

11. Wilson, C. G., Hardy, J. G., Frier, M., and Davis, S. S., Eds., *Radionuclide Imaging in Drug Research,* Croom Helm, London, 1982.
12. Frier, M., Quality control of radiopharmaceuticals, in *Radionuclide Imaging in Drug Research,* Wilson, C. G., Hardy, J. G., Frier, M., and Davis, S. S., Eds., Croom Helm, London, 1982, 75.
13. Hagan, P. L., Krejcarek, G. E., Taylor, A., and Alazraki, N., A rapid method for the labeling of albumin microspheres with In-113 and In-111: concise communication, *J. Nucl. Med.,* 19, 1055, 1978.
14. Ng, B., Shaw, S. M., Kessler, W. V., Landolt, R. R., Peck, G. E., and Dockerty, G. H., Distribution and degradation of iodine-125 albumin microspheres and technetium-99m sulfur colloid, *Can. J. Pharm. Sci.,* 15, 30, 1980.
15. Frier, M., Griffiths, P. A., and Ramsey, A. M., The physical and chemical characteristics of sulfur colloids, *Eur. J. Nucl. Med.,* 6, 255, 1981.
16. Kanke, M., Simmons, G. H., Weiss, D. L., Bivins, B. A., and DeLuca, P. P., Clearance of 141-Ce-labeled microspheres from blood and distribution in specific organs following intravenous and intraarterial administration in beagle dogs, *J. Pharm. Sci.,* 69, 755, 1980.
17. Davis, M. A. and Taube, R. A., Pulmonary perfusion imaging: acute toxicity and safety factors as a function of particle size, *J. Nucl. Med.,* 19, 1209, 1978.
18. Illum, L., Davis, S. S., Wilson, C. G., Frier, M., Hardy, J. G., and Thomas, N. W., Blood clearance and organ deposition of intravenously administered colloidal particles: the effects of particle size, nature and shape, *Int. J. Pharm.,* 12, 135, 1982.
19. Illum, L. and Davis, S. S., Specific intravenous delivery of drugs to the lungs using ion-exchange microspheres, *J. Pharm. Pharmacol.,* Suppl. 34, 89P, 1982.
20. Clarke, S. W., Pavia, D., Agnew, J. E., and Newman, S. P., Eds., Lung mucociliary clearance and the deposition of therapeutic aerosols, *Chest,* Suppl. 80, 789, 1981.
21. Newman, S. P., Pavia, D., and Clarke, S. W., Therapeutic aerosol deposition, in *Radionuclide Imaging in Drug Research,* Wilson, C. G., Hardy, J. G., Frier, M., and Davis, S. S., Eds., Croom Helm, London, 1982, 203.
22. Yeates, D. B., Warbich, A., and Aspin, A., Production of 99m-Tc labeled albumin microspheres for lung clearance studies for inhalation scanning, *Int. J. Appl. Radiat. Isot.,* 25, 578, 1974.
23. Fazio, F. and Larfortuna, C., Effect of Salbutamol on mucociliary clearance in patients with chronic bronchitis, *Chest,* 80, 827, 1981.
24. Garrard, C. S., Gerrity, T. R., Schrevier, J. F., and Yeates, D. B., The characterisation of radioaerosol deposition in the healthy lung by histogram distribution analysis, *Chest,* 80, 840, 1981.
25. Philipson, K., Radioisotope labelling for the study of lung function, *Chest,* 80, 818, 1981.
26. Cramner, P., Studies on the removal of inhaled particles by voluntary coughing, *Chest,* 80, 824, 1981.
27. Agnew, J. E., Bateman, J. R. M., Watts, M., Parmananda, V., Pavia, D., and Clarke, S. W., The importance of aerosol penetration for lung mucociliary clearance studies, *Chest,* 80, 843, 1981.
28. Malton, C. A., Hallworth, C. G., Padfield, J. M., Perkins, A., Wilson, C. G., and Davis, S. S., Deposition and clearance of inhalation aerosols in dogs and rabbits using a gamma camera, *J. Pharm. Pharmacol.,* Suppl. 34, 64P, 1982.
29. Caride, V. J., Liposomes for diagnostic imaging, in *Radiopharmaceuticals, Structure Activity Relationships,* Spencer, R. P., Ed., Grune & Stratton, New York, 1981, 477.
30. Ryder, S., Berqvist, Hagstrom, L., and Strand, S. E., Reticuloendothelial function in normal and tumor-bearing rats. Measurement with a scintillation camera technique, *Eur. J. Cancer Clin. Oncol.,* 19, 965, 1983.
31. Benacerraf, B., Sebestyen, M., and Cooper, N. S., The clearance of antigen-antibody complexes from the body by the reticuloendothelial system, *J. Immunol.,* 82, 131, 1959.
32. Kuperus, J. H., The role of phagocytosis and pinocytosis in the localization of radiotracers, in *Principles of Radiopharmacology,* Vol. 3, Colombetti, L. G., Ed., CRC Press, Boca Raton, Fla., 1979, 267.
33. Walters, M. N. I. and Papadimitriou, J. M., Phagocytosis: a review, *CRC Crit. Rev. Toxicol.,* 5, 377, 1978.
34. Reske, S. N., Vyska, K., and Feirvendegen, L. E., In vivo assessment of phagocytic properties of Kupffer cells, *J. Nucl. Med.,* 22, 405, 1981.
35. Spencer, R. P., Reticuloendothelial compounds, in *Radiopharmaceuticals, Structure Activity Relationships,* Spencer, R. P., Ed., Grune & Stratton, New York, 1981, 549.
36. Frier, M., Phagocytosis, in *Progress in Radiopharmacology,* Vol. 2, Cox, P. H., Ed., Elsevier/North Holland, Amsterdam, 1981, 249.
37. Illum, L. and Davis, S. S., Effect of non-ionic surfactants on the fate and deposition of polystyrene microspheres following intravenous administration, *J. Pharm. Sci.,* 72, 1086, 1983.
38. Illum, L. and Davis, S. S., The organ uptake of intravenously administered colloidal particles can be altered using a non-ionic surfactant (Poloxamer 338), *FEBS Lett.,* 167, 79, 1984.
39. Van Oss, C. D., Gillman, C. F., and Neuman, A. W., *Phagocytic Engulfment and Cell Adhesiveness as Cellular Surface Phenomena,* Marcel Dekker, New York, 1975.

40. Colombetti, L. G., *Biological Transport: A Historical view in Biological Transport of Radiotracers,* Colombetti, L. G. Ed., CRC Press, Boca Raton, Fla., 1982, chap. 1.
41. Atkins, H. L., Hauser, W., and Richards, P., Factors affecting the distribution of technetium sulfur colloid, *J. Reticuloendothelial Soc.,* 8, 176, 1970.
42. Scheffel, U., Rhodes, B. A., Natarajan, T. K., and Wagner, H. N., Albumin microspheres for the study of the reticuloendothelial system, *J. Nucl. Med.,* 13, 498, 1972.
43. Hinkle, G. H., Born, G. S., Kessler, W. V., and Shaw, S. M., Preferential localisation of radiolabeled liposomes in liver, *J. Pharm. Sci.,* 67, 795, 1978.
44. Jeyasingh, K., In vivo distributions of 99m-Tc-labeled liposomes, in *Radionuclide Imaging in Drug Research,* Wilson, C. G., Hardy, J. G., Frier, M., and Davis, S. S., Eds., Croom Helm, London, 1982, 243.
45. Profitt, R. T., Williams, L. E., Presant, C. A., Tin, G. W., Ohana, J. A., Gamble, R. C., and Baldeschwieler, J. D., Liposomal blockade of the reticuloendothelial system: improved tumor imaging with small unilamellar vesicles, *Science,* 220, 502, 1983.
46. Souhami, R. L., Patel, H. M., and Ryman, B. E., The effect of reticuloendothelial blockade on the blood clearance and tissue distribution of liposomes, *Biochim. Biophys. Acta,* 674, 354, 1981.
47. Oppenheim, R. C., Nanoparticles, in *Drug Delivery Systems,* Juliano, R. L., Ed., Oxford Univ. Press, New York, 1980, 177.
48. Eckelman, W. C., Meinken, G., and Richards, P., High specific activity 99m-Tc-human serum albumin, *Radiology,* 102, 185, 1977.
49. Boisvieux, J. F. and Steimer, J. L., A non-linear mathematical model for the in vivo determination of Kupffer cells number and rate of phagocytosis of radiocolloid in rats, *Int. J. Bio-Med. Comput.,* 10, 331, 1979.
50. Munz, D., Standke, R., and Horr, G., Measurement of phagocytic and proteolytic function of macrophages in liver, spleen and bone marrow, in *Progress in Radiopharmacology,* Vol. 2, Ed. Cox, P. H., Elsevier/North Holland, Amsterdam, 1981, 261.
51. Vermess, M., Doppman, J. L., Sugarbaker, P., Fisher, R. I., Chatterji, D. C., Luetzeler, J., Grimes, G., Girton, M., and Adamson, R. H., Clinical trials with a new intravenous liposoluble contrast material for computed tomography of the liver and spleen, *Radiology,* 137, 217, 1980.
52. Grimes, G., Vermess, M., Gallelli, J. F., Girton, M., and Chatterji, D. C., Formulation and evaluation of ethiodized oil emulsion for intravenous hepatography, *J. Pharm. Sci.,* 68, 52, 1979.
53. Vermess, M., Chatterji, D. C., Doppman, J. L., Grimes, G., and Adamson, R. H., Development and experimental evaluation of a contrast medium for computed tomographic examination of the liver and spleen, *J. Comput. Assist. Tomogr.,* 3, 25, 1979.
54. Lavel-Jeantet, A. M., Laval-Jeantet, M., and Bergot, C., Effect of particle size on tissue distribution of iodized emulsified fat following intravenous administration, *Invest. Radiol.,* 17, 617, 1982.
55. Davis, S. S., Purewal, T. S., Choudhury, K., Hansrani, P., Wilson, C. G., Hardy, J. G., and Frier, M., Studies on the stability and in vivo deposition of intravenous emulsions, Proc. 2nd Int. Conf. Pharm. Tech., Paris, 189, 1980.
56. Davis, S. S. and Hansrani, P. K., The evaluation of parenterally administered emulsion formulations, in *Radionuclide Imaging in Drug Research,* Wilson, C. G., Hardy, J. G., Frier, M., and Davis, S. S., Croom Helm, London, 1982, 217.
57. Davis, S. S., Emulsion systems for the delivery of drugs by the parenteral route, in *Optimisation of Drug Delivery,* Bundgaard, H., Bagger-Hansen, A., and Kofod, H., Eds., Munksgaard, Copenhagen, 1982, 198.
58. Geyer, R. P., Potential use of artificial blood substitutes, *Fed. Proc.,* 34, 1525, 1975.
59. Havron, A., Davis, M. A., Seltzer, S. E., Paskins-Hurlburt, A. J., and Hessel, S. J., Heavy-metal particulate contrast materials for computed-tomography of the liver, *J. Comput. Assist. Tomogr.,* 4, 642, 1980.
60. Cox, P. H., Ed., *Progress in Radiopharmacology,* Vol. 2, Elsevier/North Holland, Amsterdam, 1981.
61. Van Winkel, K. and Hermann, H-J., Scintigraphy of lymph nodes, *Lymphology,* 10, 107, 1977.
62. Fruhling, J., Lymph and lymphatic pathophysiology, in *Progress in Radiopharmacology,* Vol. 2, Cos, P. H., Ed., Elsevier/North Holland, Amsterdam, 1981, 223.
63. Strand, S. E., Persson, B., Christofferson, J. O., and Knoos, T., Quality control and particle size of 99m-Tc labeled colloids with gel chromatography column scanning tenhnique, in *Progress in Radiopharmacology,* Vol. 2, Cos, P. H., Ed., Elsevier/North Holland, Amsterdam, 1981, 231.
64. Ege, G. N. and Warbick, A., Lymphoscintigraphy: a comparison of 99m-Tc antimony sulfide colloid and 99m-Tc stannous phytate, *Br. J. Radiol.,* 52, 124, 1978.
65. Ege, G. N., Internal mammary lymphoscintigraphy in breast carcinoma: a study of 1072 patients, *Int. J. Radiat. Oncol. Biol. Phys.,* 2, 775, 1977.
66. Ege, G. N., Internal radiocolloid ileopelvic lymphoscintigraphy. Technique, anatomy and clinical application, *Int. J. Radiat. Oncol. Biol. Phys.* 6, 1483, 1980.

67. Aspegren, K., Strand, S. E., and Persson, B. R. R., Quantitative lymphoscintigraphy for detection of metastases to the internal mammary lymph nodes, *Acta Radiol. Oncol.*, 17, 17, 1978.
68. Richardson, V. J., Ryman, B. E., Jewkes, R. F., Jeyasingh, K., Tattershall, M. N. H., Newlands, E. S., and Kaye, S. B., Tissue distribution and tumour localisation of 99m - technetium labeled liposomes in cancer patients, *Br. J. Cancer*, 40, 35, 1979.
69. Wu, M. S., Roblins, J. C., Bugianesi, R. L., Ponpipom, M. M., and Shen, T. Y., Modified in vivo behaviour of liposomes containing synthetic glycolipids, *Biochim. Biophys. Acta*, 674, 19, 1981.
70. Bartium, R. J., Berkowitz, D. M., and Hollenberg, K., A simple radioactive microsphere method for measuring regional flow and cardiac output, *Invest. Radiol.*, 9, 126, 1974.
71. Chan, R. C., Babbs, C. F., and Vetter, R. J., Simple methods for determining the accuracy of tumor blood flow measurements using radioactive microspheres in rats, *J. Pharmacol. Meth.*, 10, 157, 1983.
72. 3M Companies, *Tracer Microspheres, A Bibliography*, 3M Co., Minneapolis, Minnesota, 1980.
73. Tepperman, B. L. and Jacobson, E. D., Measurement of gastrointestinal blood flow, *Ann. Rev. Physiol.*, 44, 71, 1982.
74. Heymann, M. A., Payne, B. D., Hoffman, J. I. E., and Rudolph, X., Blood flow measurements with radionuclide labeled particles, *Prog. Cardiovasc. Dis.*, 20, 55, 1977.
75. Segadal, L. and Svanes, K., Evaluation of the microsphere method for determination of cardiac output, *Scand. J. Clin. Lab. Invest.*, 39, 415, 1979.
76. Kirk, G. A., Adams, R., Jansen, C., and Judkins, M. P., Particulate myocardial perfusion scintigraphy: its clinical usefulness in evaluation of coronary artery disease, *Semin. Nucl. Med.*, 7, 67, 1977.
77. Hoffmann, W. E., Milekch, D. J., and Albrecht, R. F., Repeated microsphere injections in rats, *Life Sci.*, 28, 2167, 1981.
78. Slack, J. D., Kanke, M., Simmons, G. H., and DeLuca, P. P., Acute hemodynamic effects and blood pool kinetics of polystyrene microspheres following intravenous administration, *J. Pharm. Sci.*, 70, 660, 1981.
79. Freedman, S. B., Crea, F., Pratt, T. A., Brady, F., and Maseri, A., Accuracy of blood flow determinations with 99m-Tc and 113m-In labeled albumin microspheres in rabbits, *Clin. Sci.*, 65, 13, 1983.
80. Wilson, R., Shea, M., DeLandsheere, C., Deanfield, J., Brady, F., Turton, D., Selwyn, A., and Maseri, A., A new method for in vivo quantitation of regional myocardial blood flow using C-11 labeled micropheres and positron tomography, *Eur. Heart J.*, 4, 76, 1983.
81. Madden, R. B., Agostino, D., and Gyure, L., Some unusual aspects of microsphere migration at the microcirculatory level, *Arch. Surg.*, 101, 425, 1970.
82. Davis, S. S., The use of scintigraphic methods for the evaluation of drug dosage forms in the gastrointestinal tract, in *Topics in Pharmaceutical Sciences*, Breimer, D. D. and Speiser, P. P., Eds., Elsevier, Amsterdam, 1983, 205.
83. Digenis, G. A., The utilization of short lived radionuclides in the assessment of formulation and in vivo deposition of drugs, in *Radionuclide Imaging in Drug Research*, Wilson, C. G., Hardy, J. G., Frier, M., and Davis, S. S., Eds., Croom Helm, London, 1982, 103.
84. Digenis, G. A., Beihn, R. M., Theodorakis, M. C., and Shambhu, M. B., Use of 99m-Tc labeled triethylene-tetramine-polystyrene resin for measuring the gastric emptying rate in humans, *J. Pharm. Sci.*, 66, 442, 1977.
85. Wirth, N., Swanson, D., Shapiro, B., Nakajo, M., Coffey, J. L., Eckhauser, F., and Owyang, C., A conveniently prepared Tc-99m resin for semisolid gastric emptying studies, *J. Nucl. Med.*, 24, 511, 1983.
86. Theodorakis, M. C., Groutas, W. C., Whitlock, T. W., and Tran, K., Tc-99m labeled polystyrene and cellulose macromolecules: agents for gastrointestinal scintigraphy, *J. Nucl. Med*, 23, 693, 1982.
87. Theodorakis, M. C., Digenis, G. A., Beihn, R. M., Shambhu, M. B., and DeLand, F. H., Rate and pattern of gastric emptying in humans using 99m-Tc-labeled triethylene-tetramine-polystyrene resin, *J. Pharm. Sci.*, 69, 568, 1980.
88. Hinder, R. A. and Kelly, K. A., Canine gastric emptying of solids and liquids, *Am. J. Physiol.*, 233, E335, 1977.
89. Palin, K. J., Whalley, D. R., Wilson, C. G., Phillips, A. J., and Davis, S. S., Gastric emptying of oils in the rat, in *Radionuclide Imaging in Drug Research*, Wilson, C. G., Hardy, J. G., Frier, M., and Davis, S. S., Eds., Croom Helm, London, 1982, 315.
90. Palin, K. J., Whalley, D. R., Wilson, C. G., Davis, S. S., and Phillips, A. J., Determination of gastric emptying profiles in the rat: influence of oil structure and volume, *Int. J. Pharm.*, 12, 315, 1982.
91. Groves, M., Particle size characterisation in dispersions, *J. Dispersion Sci. Technol.*, 1, 97, 1980.
92. Berne, B. J. and Pecora, B., *Dynamic Light Scattering with Applications to Chemistry, Biology and Physics*, Plenum Press, New York, 1977.
93. Lim, T. K., Bloomfield, V. A., and Krejcarek, C., Size and charge distribution of radiocolloid particles, *Intern. J. Appl. Radiat. Isotop.*, 30, 531, 1979.

94. Ryman, B. E. and Tattershall, M. H. N., Possible tumor localisation of Tc-99m-labeled liposomes: effects of lipid composition, charge and liposome size, *J. Nucl. Med.*, 19, 1049, 1978.
95. Seno, S., Tanaka, A., Urata, M., Hirata, K., Nakatsuka, H., and Yamamoto, S., Phagocytic response of rat liver capillary endothelial cells and Kupffer cells to positive and negative charged iron colloid particles, *Cell Struct. Funct.*, 1, 119, 1975.
96. Caride, V. J. and Zaret, B. L., Liposome accumulation in regions of experimental myocardial infarction, *Science*, 198, 735, 1977.
97. Wilkins, D. J. and Myers, P. A., Studies on the relationship between the electrophoretic properties of colloids and their blood clearance and organ distribution in the rat, *Br. J. Exp. Pathol.*, 47, 568, 1966.
98. Bangham, A. D., Flemans, R., Heard, D. H., and Seaman, G. V. F., An apparatus for microelectrophoresis of small particles, *Nature (London)* 182, 642, 1958.
99. Preece, A. W. and Luckman, N. P., A laser Doppler cytopherometer for measurement of electrophoretic mobility of bioparticles, *Phys. Med. Biol.*, 26, 11, 1981.
100. Andreade, J. D., Ma, S. M., King, R. N., and Gregonis, D. E., Contact angles at the solid-water interface, *J. Colloid Interface Sci.*, 72, 488, 1979.
101. Malmqvist, T., Bacterial hydrophobicity measured as partition of palmitic acid between two immiscible phases of cell surface and buffer, *Acta Pathol. Microbiol. Immunol. Scand.*, Sect. B, 91, 69, 1983.
102. Magnusson, K. E. and Johansson, G., Probing the surface of Salmonella typhimurium and Salmonella minnesota and R bacteria by aqueous biphasic partitioning in systems containing hydrophobic and charged polymers, *FEMS Microbiol., Lett.*, 2, 225, 1977.
103. Walter, H., Partition of cells in two polymer aqueous phases: a surface affinity method for cell separation, *Meth. Cell Sep.*, 1, 307, 1977.
104. Halperin, G. and Shaltiel, S., Homologous series of alkyl agaroses discriminate between erythrocytes from different sources, *Biochem. Biophys. Res. Commun.*, 72, 1497, 1976.
105. Magnusson, K. E., The hydrophobic effect and how it can be measured with relevance for cell-cell interactions, *Scand. J. Inject. Dis. Suppl.*, 24, 131, 1980.
106. Green, D. E. and Zande, H. V., The determinants of the changes in fluorescence of 8-aniline-1-naphthalene sulfonic acid in particle systems, *Biochem. Biophys. Res. Commun.*, 107, 1300, 1982.
107. Norde, W., Adsorption of proteins at solid surfaces, in *Adhesion and Adsorption of Polymers*, Lee, H-L., Ed., Plenum Press, New York, 1979, 801.
108. Gudewicz, P. W., Molnar, J., Siefring, J., Credo, B., and Lorand, L., Factors regulating gelatinized iodine-125 latex particle uptake by mouse peritoneal macrophage monolayers, *Res. J. Reticuloend. Soc.*, Suppl. 24, 15A, 1979.
109. Scherphof, G., Roerdink, F., Dijkstra, J., Ellens, H., Zanger, R., and Wisse, E., Uptake of liposomes by rat and mouse hepatocytes and Kupffer cells, *Biol. Cell*, 47, 47, 1983.
110. Davis, M. A., Jones, A. G., and Trindade, H., A rapid and accurate method for sizing radiocolloids, *J. Nucl. Med.*, 15, 923, 1974.
111. Frier, M., Griffiths, P. A., and Ramsey, A. M., The biological fate of sulfur colloid, *Eur. J. Nucl. Med.*, 6, 371, 1981.
112. Garzon, O. L., Palcos de Enquin, M. C., and Radicella, R., 99-Tc^{m}-labeled colloid, *Int. J. Appl. Radiat. Isot.*, 16, 613, 1965.
113. Warbick, A., Ege, G. N., Henkelman, R. M., Maier, G., and Lyster, D. M., An evaluation of radiocolloid sizing techniques, *J. Nucl. Med.*, 18, 827, 1977.
114. Ramsey, A. M., personal communication, 1981.
115. Huh, Y., Donaldson, G. W. M., and Johnston, F. J., A radiation-induced bonding of iodine at the surface of uniform polystyrene particles, *Radiat. Res.*, 60, 42, 1974.

INDEX

A

B

C

D

E

F

G

H

I

K

L

M

N

O

P

Q

R

S

T

U

V

W

Z